W9-BEC-472

The Time of Our Lives

The Time of Our Lives

A Critical History of Temporality

David Couzens Hoy

The MIT Press
Cambridge, Massachusetts
London, England

For information about special quantity discounts, please email special_sales@ mitpress.mit.edu

This book was set in Syntax and Times New Roman by SNP Best-set Typesetter Ltd., Hong Kong. Printed and bound in the United States of America.

Library of Congress Cataloging-in-Publication Data

Hoy, David Couzens.
The time of our lives: a critical history of temporality / David Couzens Hoy.
    p.   cm.
Includes bibliographical references and index.
ISBN 978-0-262-01304-5 (hc. : alk. paper)
1. Time—Philosophy. I. Title.
BD638.H585   2009
115—dc22

                                                        2008035983

10 9 8 7 6 5 4 3 2 1

Contents

Preface

This book is the first volume in a planned two-volume study of the history of consciousness. This volume represents a history of time-consciousness. The next volume, currently in progress, focuses on the history of self-consciousness. This order is itself a philosophical problem and it involves some crucial philosophical decisions. Some philosophers would expect the study of self-consciousness to come before the study of time-consciousness. These philosophers have intuitions formed by the Kantian and neo-Kantian tradition. According to this tradition, time is a form of intuition and is imposed by the mind on experience. Holders of this view might well expect, then, a theory of self-consciousness to come before (both logically and temporally) a theory of time-consciousness.

By starting with time-consciousness, this book challenges the logical ordering that puts mind before time. The thought that is being explored in the phenomenological tradition is that temporality is a condition for the possibility of subjectivity. The assumption that the reverse is the case must therefore not be taken for granted. Along the way, however, several other aspects of the Kantian tradition are also called into question. Among them is the very idea of something "coming before" something else. The transcendental program of showing the logically prior "conditions for the possibility of experience in general" is challenged here. Simply reversing the ordering of the relation of mind and time would not break with transcendental philosophy. To make that break, a thoroughly pragmatic or hermeneutical philosophy will have to give up the project of explaining which is the more primordial, mind or time, and which is derived. Furthermore, the very concepts, *mind* and *time*, must be problematized. Although they are not necessarily abandoned, the extent to which they surreptitiously carry with them much philosophical baggage should become clearer as this historical study of time- and self-consciousness unfolds. In this volume, the idea of time-consciousness itself is called into question right at the beginning. Whether it survives at the end or not, it undergoes conceptual transformations that might well make it unrecognizable to its most famous proponent, Edmund Husserl.

A subsidiary thesis of this book is that the history of philosophy can make a philosophical difference. The method of *critical history*, or *genealogy*, is intended to challenge predominant understandings of what the philosophical issues are supposed to be by shaking the foundations of philosophy and showing that philosophical concepts and issues are not fixed in stone forever. The thought that there are perennial problems of philosophy that have not changed is thus itself to be questioned. What philosophy itself is concerned with and how it has changed needs to be shown by a critical history of philosophical themes. This history has the potential to reveal and perhaps even to cause meaning changes, conceptual shifts, and even tectonic transformations in the overall philosophical landscape. If these studies contribute to those transformations even to a small extent, they will have served their purpose.

Acknowledgments

This first volume has been a long time in gestation, and its provenance is indebted to many people. First and foremost, I should acknowledge the metaphilosophical influence of my close friend, Richard Rorty, the time of whose life sadly came to an end on June 8, 2007, the day that this manuscript was completed in Paris. If I believed in a permanent philosophical pantheon, I would certainly argue for his preeminence in it. Although he did not thematize temporality, the ideas in this book would not have been possible without our many conversations.

Next I wish to acknowledge in-depth conversations with Burkhard Kümmerer, professor of mathematics and physics in Darmstadt, Germany. Although we have known each other since 1968, in the last decade he has helped me to understand much about the time of physics, even if this book is not about physical time as such. A gifted teacher, he made many things clear to me in ways that prevented major errors I might otherwise have made. Of course, errors that I do make are entirely on my own conscience. In any case, I hereby dedicate this book to the entire Kümmerer family—Ursula, Dorothea, Burkhard, Andrea, Matthias, Henrike, and Frieder—and in particular, to the memory of Dr. Emil Kümmerer.

I should also express my gratitude to the neuroscience faculty and staff of Stanford University, including Dr. Helen Brontë-

Stewart, Dr. Jaimie Henderson, Dr. Gary Heit, Wendy Cole, Kay McGuire, and all the members of their teams. Their combined efforts have added greatly to the quality not only of my life, but the lives of many others as well. Without their expertise, as well as that of Dr. Cathleen Miller and Dr. Josh Novic, there are three of my recent books that could not have been written: one in the past (*Critical Resistance*), the present one on the history of time-consciousness, and the history of self-consciousness that should appear in the near future.

My graduate students and academic colleagues in the departments of philosophy, history of consciousness, politics, anthropology, and literature at the University of California, Santa Cruz, have been a continual source of stimulation. Research on this project has been supported by funds that came with my appointments as Presidential Chair in Philosophy as well as Distinguished Professor of Humanities. Additional funding came through the Santa Cruz Institute for Humanities Research and the Committee on Research. MIT Press editor Tom Stone as well as his helpful readers were also crucially involved in the final transformations of the manuscript into the book.

Most of all, I acknowledge Jocelyn and Meredith, for they are the ones who have made and who continue to make the time of my own life both meaningful and beautiful.

Introduction

In contrast to the exquisite inquiries of Marcel Proust into how time is experienced, philosophical attempts to describe lived temporality may appear graceless. Nevertheless, there is an appealing aesthetic quality and even a certain beauty in the subtleties, distinctions, and intricacies of the great philosophers as they work on an intractable problem such as time. Proust's goal is not so different from philosophers such as Kant, Nietzsche, Heidegger, Bergson, and Deleuze, who want to identify the source of time. Starting from the recognition of the increasingly rapid loss of time, the task becomes to explain what it is that we lose as time goes by. From these explanations comes a hope to recover or regain, not the time we have lost, but the time that remains: the time of our lives.

The project of all philosophy may be to gain reconciliation with time, whether or not a particular philosopher includes an explicit analysis of time. Not every philosopher has made time an express topic, however, and this study engages only a further subset of those who did. In particular, this study focuses on the tradition of phenomenology with attention as well to some precursors and successors. The purpose is to see how phenomenological philosophers have tried to locate the source of time, how they analyze time's passing, and finally, again like Proust, how they depict our relation to time once it has been regained. Resentment of mortality and reconciliation with finitude are equally possible reactions to time's passing. The question becomes the normative one of how best to relate to time. There is also the political question of the optimal strategy for dealing with time's passing on the level of the social and historical. Nostalgia for the past and hope for the future each have their adherents, for instance. Yet there are those who reject both of these attitudes. If we give up utopian hopes, however, are we then simply resigned to the temporal finitude that eats away at our lives? Or in the manner of Proust or Nietzsche, can we become reconciled to time by creating our lives all over again and turning life into literature?

These questions should indicate that this book is not primarily about the nature of time in general. The focus is instead on the history of the phenomenology of time as time shows up in human lives. To write about the nature of time in and of itself would require an exploration of a complex array of issues about the status of what could be called "scientific" or "objective" or "universal" time, that is to say, the "time of the universe." Restricting the book to the phenomenology of human temporality—to "the time of our lives"—raises an equally formidable but different set of questions. In this book some of the questions raised by our authors are the following. Is the time of our lives a function of a life as a whole, a lifetime, or can it be condensed into a single moment of vision? Does a life have a unity that runs through it, or is the unity of time, and of a life, a narrative, a story, a fiction, or even an illusion? Can time be perceived? What is the time like that we encounter in our experience of our world and ourselves? Is the time of our lives the same as the time of nature or of history? In particular, if time runs through our lives, in which direction does it run? Does time come toward us from the future, as Martin Heidegger maintained, from behind us through the past, as Pierre Bourdieu asserted, or from the present, cycling perhaps in an eternal recurrence, as Friedrich Nietzsche speculated? Then there is Immanuel Kant's question: is temporality a feature of us or of the world? That is, is the time of our lives subjective or objective, or is there a third possibility?

Such questions could well require much more than one lifetime to answer. When they are approached from the human or phenomenological viewpoint rather than from the standpoint of physics or metaphysics, however, the questions take on a different and more accessible character. To pick up on the last one as an example, a major issue is whether the time of our lives is in fact merely a subjective or perceptual phenomenon, or whether it is just as real as the time of the universe. One might think that making that distinction into a sharp difference in kind solves the problems by differentiating between, say, the way psychology might deal with time and the way physics postulates time. On this approach, physical time will be taken as real and psychological time will be construed as unreal, as a merely subjective illusion. For phenomenology, however, the very distinction between the subjective and the objective, between the physical and the psychological, is what is at issue.

To avoid ambiguous references to "time," where whether one is talking about universal time or human time is unclear, let me stipulate provisionally a conceptual distinction between the terms "time" and "temporality." The term "time" can be used to refer to universal time, clock time, or objective time. In contrast, "temporality" is time insofar as it manifests itself in human existence. Note that I have cautiously not specified temporality as "subjective time," or "experienced time," because these terms are at issue. Instead, my intention is to discuss philosophical accounts of what has been called "lived time," or "human temporality"—hence, "the time of our lives." Because our philosophers often do not make this distinction between the time of the universe and the time of our lives, it will be hard to maintain in every instance. We may have to ask on occasion, what "time" is it? Nevertheless, the distinction will be useful for demarcating and delimiting the issues of this study.

In the history of phenomenology, not attending to this distinction has led to some philosophical labyrinths. For instance, the first self-described phenomenologist, Edmund Husserl (1859–1938), was never able to complete to his own satisfaction his book, *Zur Phänomenologie des inneren Zeitbewusstseins* (best translated as *On the Phenomenology of Internal Time Consciousness*). Right away, note the ambiguity that is caused by this term "internal." Is it time or consciousness that is "internal"? The term "internal" represents a philosophical fixation on the Kantian question of whether time is real, or whether it is imposed by the mind on the world. The terms "internal" and "consciousness" also suggest that time is a thing or a quality imposed by a "subject" on "experience." All these terms are problematic and should be used with care.

The initial task of this book will be to explain how the "time of our lives" emerges as a separate problem from the "time of the universe." This book is intended as an introduction that explains how the problems shift when viewed from the distinctive point of view of temporality as a problem for our lives. If this book is an introduction, it is not necessarily introductory. The issues are complex and the existential perspective emerges only gradually from a historical discussion where philosophers had other goals as well. Some were trying to say what time really was, or whether it was real at all. That is not the problem here, because temporality must be *experienced* as real. Others were trying to describe temporality from a subjective as opposed to the objective point of view. That enterprise comes closer to the project here, but it is not exactly the same because for other philosophers such as Heidegger and Merleau-Ponty, the subject–object distinction is what is in question. Heidegger and Merleau-Ponty want to explain that distinction as an *emergent* one that grows out of primordial temporality and that therefore cannot be used to determine the status of temporality beforehand.

As the book progresses and the phenomenological issues get sorted out, the questions become more explicitly social and political or "critical." They also become more personal or "existential." So as the history of the concept of temporality is reviewed, the standpoint of this book will also emerge. In Ludwig Wittgenstein's words, "Light dawns gradually over the whole."[1] In plainer language, the book will discuss the history of the phenomenological concept of time-consciousness to the extent that it relates to wider human interests and purposes involved in questions about temporality, or the passing of time.[2]

Temporality

To survey the landscape of the book in more detail, I will now explain the concept of temporality in a preliminary fashion, as well as the method of analysis. For Kant, as chapter 1 shows, the main question about time is whether the time of what he calls "the starry skies above" is objective or subjective, that is, mind independent or mind dependent, real or ideal. The same question could not be asked about what I am calling temporality. Clearly human experience is temporal, whether or not we are conscious of the temporal. Also, it seems hard to deny that we can be conscious of the temporality of existence. We know, then, that temporality is real. The question of the source of time, that is, of whether time comes from the mind or the world, is obviated by the undeniable occurrence of temporality. The reality of temporality seems equally objective and subjective. Standard parlance would say that we recognize that the experience of the flow going faster or slower is subjective, yet nevertheless it is generally acknowledged that the flow is objectively happening. So the character of temporality—for example, whether it goes by quickly or slowly—appears to be dependent on the mind and would thus be said to be subjective. It is hard to deny that time goes by, however, and thus it seems incontestable that we are experiencing a phenomenon that is genuinely objective.

Focusing on temporality allows the phenomenologist to avoid many of the metaphysical questions that arise about the reality or the ideality of time. Other philosophical issues are not so easily dispelled, however. To return to the question, for instance, about whether time can be perceived, I note that we perceive ourselves as in time, and we perceive temporal sequence. We even perceive temporality insofar as we have the experience of time passing. Thus, we can say that time passes quickly or slowly, that a piece of music was played allegro or adagio. But do we perceive time itself? It is hard to know what there would be to perceive. The steady advance of the second hand? Punching into the atomic clock? These are temporal phenomena, but they are not time.

This account involves the famous problem known as the arrow of time, which I have already invoked as a question about objective time. When temporality is what is at stake, the question becomes more particularly, in which direction do we experience the flow of temporality? Is it experienced, for instance, as coming from the past into the present and then flowing on into the future? Or does it come out of the future into the present and then on into the past? We can even ask whether the fluvial metaphor makes any sense at all. Water flows relative to the banks of the river, but relative to what could temporality be said to flow?

A related conundrum concerns the size of the present. Is it just an infinitesimal blip between the past and the future? If this were the case, and if the past and the future do not exist, then what does exist is certainly very fleeting. If the present is not to disappear, it must be more than the minuscule gap between the moment that just was and the moment that is about to come.

The discussion has now turned to the issue of the *oneness* of temporality at any given point in time. It also leads to the issue of the *unity* of temporality *over time*, which is not the same. The notions of *oneness* and *unity* are usefully distinguished. The problem about *oneness* concerns the question of how we know that any given moment is the same for everyone. That is, how is clock time possible? We can say, "Synchronize your watches," but this presupposes public time, the source of which is supposedly *objective* time. But that term is precisely the problem. The question about *unity* arises from asking about the cohesiveness of temporality. Heidegger, for instance, wanted to know about the temporal connection of a life between birth and death. The connectedness of our lives over time is thus a central issue in our ability to be authentic beings insofar as inauthenticity is precisely the lack of temporal unity.

Genealogy and Phenomenology

These are some questions about time that turn into questions about temporality. This book will test the value of distinguishing, at least conceptually, time and temporality. A conceptual distinction is not necessarily a distinction that can be made in experience. Kant distinguishes concepts and intuitions, for instance, but he does not claim that this distinction can be experienced. Instead, every experience must combine concepts and intuitions. Specific chapters will focus on the questions about temporality that are within the purview of phenomenology. Insofar as this book represents a history of the concept of temporality, it can be read as an introduction to the philosophical issues. At the same time, however, it must enter into debate with the phenomenologists about temporal experience. It is, therefore, a *critical* history of temporality. The term "critical" here implies a connection with the tradition of critical theory. The allegiances of this book are less with the Frankfurt School, however, than with Michel Foucault's use of the genealogical method. In Foucault's genealogy of ethics, for instance, he is writing not about the explicit moral rules that people espouse, but more about the underlying *ethos*, or "ethical substance" of different cultures, whether ancient or modern, Western or Eastern. Ethical substance, a term he borrows without acknowledgment from Hegel's *Phenomenology of Spirit*, includes the basic *ethos* of a culture's ethical formation. This *ethos* helps to explain why people adhere to ethical norms, and what they hope to become through their pursuit of these norms.

In a study of temporality, the corollary of the *ethos* is the sense of time passing and the strategies that emerge for dealing with it.

This book therefore supplements the history of phenomenology in chapters 1 through 4 with a genealogical account of the relevance of this history for contemporary life in chapter 5. As the historical account of the phenomenology of the present, the past, and the future progresses, the normative issues about temporality will begin to appear, until finally in chapter 5, the question of how to reconcile ourselves to the passing of our lives is addressed directly.

In sum, this book is a selective study of the history of modern continental philosophy with particular attention to accounts of the temporality of the present, past, and future. The book differs from others in that it devotes a chapter to each of these three modes of time, and discusses the phenomenological philosophers who had the most to say about each modality. Insofar as it is difficult to keep the temporal dimensions entirely separate, it will be necessary to refer to the other two dimensions in discussing one. I emphasize the philosophers who have the most to say about the thematic problems associated with the particular mode of temporality over those who have less to say about it.

As that organization has disadvantages for a reader who is more interested in a particular thinker than in the separate modes of present, past, and future, I wish to point out that the book can be read either horizontally or vertically. By that I mean that if a reader were particularly interested in Heidegger or in poststructuralist philosophers such as Derrida or Deleuze, it would be possible to read the sections in different chapters bearing on that philosopher.[3]

In contrast to that horizontal way of working through the chapters, is the standard, vertical way of reading each chapter at a time, with its topical focus on problems arising from a particular dimension of temporality. What follows is a brief indication of the philosophical issues discussed in each chapter.

Chapter 1 sets up the issue about the source of time through an account of Kant's interpretation of time and Heidegger's deliberate misreading of Kant. One important point that emerges from this comparison is that there is a significant difference between the Kantian approach to temporality through "faculty psychology," and the phenomenological approach through "duration." Although I read Husserl as a theorist of duration, I find elements of both duration and faculty psychology in his student Martin Heidegger. Issues about normativity come up with the question of whether Heidegger's distinction between the authentic and the inauthentic is a moral distinction. Heidegger denies that the authentic–inauthentic distinction is value-laden, but I maintain that it has to be understood at least as the source of values, that is, as the basis of normativity. Other issues include a discussion of whether the mind is the source of temporality or, if that thought is not surprising enough, whether temporality could be the source of subjectivity. Questions in the philosophy of mind come up in discussing the tensions between Kant's and Heidegger's notions of subjectivity. Furthermore, attention has to be given to how to account for the synchronic oneness of temporality at any given moment as well as the diachronic unity of temporality over time. Chapter 1 is intended for readers with particular interests in Kant and Heidegger. The general reader may wish to start instead with chapter 2, perhaps coming back to chapter 1 later.

Chapters 2, 3, and 4 focus respectively on each of the dimensions of present, past, and future. Chapter 2 raises the question, what is the present? The discussion starts with Hegel and William James before turning to the phenomenologists proper, namely, Husserl, the early Heidegger, and Merleau-Ponty. Hegel and James bring out problems in specifying exactly what "now" means. Hegel raises the question of whether the word "now" works as an indexical or a universal. James sees the present as ambiguous between instantaneity and duration. Husserl's theory of internal time-consciousness suggests how it can be both. Because Merleau-Ponty is an influential interpreter of Husserl, the order of exposition puts Merleau-Ponty before Heidegger. Merleau-Ponty sees the source of the present in each individual, and despite his account of inter-subjectivity, the problem is whether he can escape the quagmire of temporal idealism. Heidegger distinguishes various kinds of temporality, with different evaluations of the significance of the present. The question also becomes whether the emphasis on the temporal present makes Merleau-Ponty's view susceptible to Derrida's critique of the metaphysics of presence. Derrida's famous critique of presence is both derived from and applied to Heidegger. Another source for it is Nietzsche's doctrine of eternal return, which is examined to see whether it can validate the primacy of the present. Along the way, the chapter also explores the limitations of two common metaphors for time—time as a river and time as a string of pearls—when these are applied not to time, but instead only to temporality. Merleau-Ponty's images of the fountain and the railroad car are explored as alternative metaphors for temporality.

The chapter on the past, chapter 3, is concerned with issues about where time goes and whether the past can be changed. Brief lessons are extracted from the German tradition, including Husserl, Heidegger, and Gadamer, as well as the French tradition, including Jean-Paul Sartre, Pierre Bourdieu, and Michel Foucault. Then the discussion turns to Henri Bergson, as interpreted first by Maurice Merleau-Ponty and then by Gilles Deleuze. Despite Bergson's problematic encounter with Einstein and relativity theory, Bergson's account of duration as an expandable cone is significantly different in philosophically interesting ways from Husserl's graph, which still represents temporality as linear and punctual.

Chapter 4 concentrates on the future, raising issues about the phenomenology of the futural, but also about political implications. In particular, the question is whether we need to hope for future utopias in order to justify present actions. Action requires a sense of direction, which has been imperiled by the speed of modern life and the need to act without reflection. The models for a historical sense of hope are Kant and Hegel, who are then contrasted with philosophers who do not share the hopes of the Enlightenment. In contrast to Marx's hope for Revolution as a response to temporality, I consider Žižek's attitude of Refusal, in the manner of Bartleby, as well as Derrida's "roguish" political program of deconstructive genealogy. The genealogical dimension of this study starts to become evident in this chapter, and it appears explicitly with the normative issues raised in the next chapter.

Chapter 5 thus concludes with some existential strategies for dealing with the apparent flow of temporality. Proust and Benjamin are contrasted on the effectiveness of reminiscence and remembrance for ameliorating the sting of time's passing. In addition to a discussion of Heidegger's political attitudes and the changes in his thought, this chapter takes seriously Slavoj Žižek's critique of both Heidegger and poststructuralism. Finally, it concludes with a reading of Deleuze that links without synthesizing Husserl's and Bergson's approaches to temporality. A postscript on the genealogical method in contrast to phenomenology and critical theory clarifies the philosophical allegiances of this study of the time of our lives.

1 In Search of Lost Time: Kant and Heidegger

Where should a history of the phenomenology of temporality begin? Strictly speaking, phenomenology in the distinctive sense that it has today starts with Edmund Husserl. Martin Heidegger and Maurice Merleau-Ponty are then among those who subsequently self-identified as phenomenologists, although Heidegger's connection to Husserl makes that label problematic. Any such history would have to recognize, however, that phenomenology emerges from a longer and wider tradition that includes major figures such as Immanuel Kant as well as Husserl's precursors and near contemporaries such as William James or Franz Brentano.

This chapter begins accordingly with an introductory account of Kant in the first section, followed by a discussion of Heidegger's reading of Kant in the second section, and of the development of the early Heidegger's own efforts at explaining temporality in the third section. In the broadest terms, the principal thread is the search for the source of temporality. Although vastly different in style from Proust's project of searching for lost time, the philosophical search for the source of time is similar in its goals. Proust's project is informed, after all, by Bergson's theory of temporality, as we will see in later chapters. The question raised by both literature and philosophy concerns time's passing, and how to reconcile ourselves to it. The philosophical project is to construct a theory that recognizes temporality as an unavoidable feature of experience. What must be explained is our sense both that time is independent of us and that our experience introduces qualitative elements into the experience of temporality.

In more technical terms, the question in the Kantian tradition is whether time is mind dependent or mind independent. Kant seems to have wanted to have it both ways, so the question in the first section is whether he succeeded. In the second section we will discuss Heidegger's interpretation whereby Kant missed his own cue when in rewriting the first *Critique* Kant played down the role of the imagination in the production of temporal experience. In the third section Heidegger's own analysis of the temporality of phenomena such as joy, anxiety, and boredom is examined to see how he argues for his inversion of the problem. On his account, the question is not whether time is mind dependent or mind independent, but whether mind is dependent on or independent of a prior temporalization of the world.

The purpose of these accounts is not to explore all the complexities of Kant and Heidegger scholarship, but to highlight what is involved in the project of searching for the source of time. Later chapters go into detail about how these and other philosophers viewed the different dimensions of time—present, past, and future. These opening accounts of Kant and Heidegger are intended to provide a framework for the subsequent investigations of these three dimensions of temporality and the particular problems that go along with each of them.

Kant on the Source of Time

What is the source of time? If that seems like a strange question, try thinking about what the source of temporality might be. Consider the distinction that I stipulated in the introduction between the time of the universe as opposed to the temporality of our lives. Given the question about the source of temporality, this distinction between objective time and lived temporality implies that there are only two possibilities for the source of time, the world or ourselves. If temporality is the time of our lives, as opposed to the time of the universe, then a plausible answer is that temporality comes from us, unlike time, which must come from the universe. Philosophy is not so easily satisfied, however, by such a quick answer to the question. Philosophical conscience forces a further question: what is meant by "comes from us"? This question in turn divides into two others: (1) who is this "we"? and (2) what does "comes from" mean?

Kant and Heidegger are two philosophers who answer these questions differently, despite Heidegger's attempt to elicit his own view from Kant. Although Kant criticizes Descartes for starting from the "I think" or the cogito, Kant himself reduces all that "we" are empirically to a transcendental "I," which he calls the "transcendental unity of apperception." This unity is the purely formal principle of the identity of experience and is completely empty of content. Why is unity so important, then, and what is its relevance to Kant's explanation of time? Kant's method of explanation of the genesis of experience is called "faculty psychology." If at first glance it does not seem promising to maintain that this transcendental unity of apperception could be the source of time, nevertheless, Kant does entertain the thought that the mind is the source of time, as I will now explain.

Kant's faculty psychology is the precursor of modern cognitive science insofar as he is the first philosopher to use a computational model to explain the mind's production of experience.[1] In this type of explanation, the mind is not a tabula rasa, an empty slate, or a black box, as it is for the empiricists. On the empiricists' model, the mind's reception of data is already experienced. For Kant, in contrast, experience is the output of a complicated prior process of "synthesis," which produces experience but is not itself experienced. The input, which also is not experienced as such, he calls intuition, and it "comes from" the faculty of sensibility. At first this input is an undigested multiplicity of sensations. This input must therefore be "unified" or "synthesized" in Kantian terminology, or "processed" in more recent terminology, by being brought under concepts supplied by the faculty of the understanding. Thus, the data come from the world and the concepts from the mind. These are the only two possibilities, and Kant maintains that concepts without intuitions are empty, and intuitions without concepts are blind. For Kant—who does not yet distinguish time from temporality in the manner of the later phenomenologists—the source of time must be either the world or the mind. That is to say, time must be either real or ideal. It must be either mind independent or mind dependent.

Which is it? Kant's answer is that time is not a concept, but neither is it the *content* of an intuition. Instead, he calls it a *form* of intuition. If the only two possibilities are concepts and intuitions, what does he mean by this idea of form? In the "Transcendental Aesthetic" of the first *Critique* he offers some arguments for why time is not a concept and why it is also not an intuited content. Time is not the content of intuition because time is (a) not empirical, and (b) necessary. That time is not empirical means that it cannot be perceived. That time is necessary means that although there could be time in the absence of appearances, there could not be any appearances without time. Necessity, furthermore, cannot be determined from empirical matters only, but is contributed by the mind. Time is not a concept primarily because it is a unitary phenomenon, which Kant calls a singularity, since the parts of time are all in one time. Insofar as concepts capture only generalities, not singularities, time then cannot be a concept. Kant decides to call time a *form* of intuition because all experiences are temporal (determined as successive in time), even if only some experiences involve time directly.

As a response to the question of where time comes from, the answer that time is a form of intuition might appear to be trying to have it both ways.[2] On the one hand, insofar as time is a *form* of intuition, it comes from the mind. On the other hand, however, insofar as it is a form of *intuition,* and intuition receives data from the real world, time is empirically real. Kant is thus in some respects an idealist about time insofar as he claims that time is mind dependent, and in other respects he is not an idealist. He maintains that he can be both an *empirical* realist about time, insofar as he regards time as independently real, and a *transcendental* idealist about time, insofar as he regards time as ultimately mind dependent.

What "idealism" means in the Kantian framework is obviously quite complex. Kant mentions several kinds of idealism in the *Critique of Pure Reason*, including his own transcendental idealism. Kant's critique of Descartes in the "Refutation of Idealism" depends on viewing Descartes as what Kant calls a *problematic* idealist. Unlike the *dogmatic* idealist, Berkeley, who *denies* the existence of objects in space, the problematic idealist merely *doubts* their existence. Kant's strategy is then to turn the tables on this version of empirical idealism by proving that "even our *inner experience*, undoubted by Descartes, is possible only under the presupposition of outer experience" (B275). Kant believes that it is a "scandal of philosophy and universal human reason" that we lack a proof of the external world.[3]

Indeed, there is a question about what is even meant by common terms like "external" or images of the "things outside us." Such phrases could mean experience of "objects as outer" or they could mean, more strongly, experience of "mind-independent things." Hallucinations, for instance, are cases of the former but not of the latter. What has to be proved is that there is input and that experience is not coming from me alone. Thus, even if I am a brain in a vat and am deceptively programmed by an evil genius with false input, there must be (1) external input (even if it is illusory), and (2) objective orderability. Inner experience is orderable (determined in time) only if an outer *order* is being experienced.[4] If experience were completely chaotic, I could not distinguish inner and outer, and I probably could not talk about an "I" at all. Even an experience of the inner, such as a hallucination, is objective in the sense that it must be orderable as internal. I say of a particular experience that it is just a hallucination and only inner, because I know that it is not orderable along with outer experience. That is why people can know that they are having hallucinations. They can know that they are hallucinating because at some level they intuit that these experiences could not be externally real.[5]

If one were to ask which experiences are really outside, the answer would depend on what "outside" means. On Kant's account, objectivity implies orderability, where orderability is time-determination. What this means is that for Kant the outside is determined not by direct perception but by application of the rules of experience. The rules, and not some manner of introspecting the phenomenon, determine what counts as being outside and what does not. Kant also believes, however, that the representation of something persistent is not a persisting representation. Whereas the former is invariably fleeting and changeable, it necessarily refers to something that persists without being represented itself. In the section of the *Critique of Pure Reason* called the First Analogy, Kant identifies the source of this persistence as substance. Let me review Kant's argument for persisting substance with the purpose of eliciting what that argument tells us further about his conception of time.

Berkeley maintained that to be is to be perceived. But, says Kant in the first *Critique* (B225), time cannot be perceived. Does that entail, then, that time does not exist? The answer is no: the nonexistence of time does not follow from our inability to perceive it. For Kant, the main reason time cannot be perceived is because although perception is constantly changing, time itself does not change. Time is the framework for all perception, or more precisely, the condition for the perception of any object whatsoever, including temporality.

This argument represents a revolutionary perspective on time. Instead of talking about the nature of time as it is in itself, Kant focuses attention on time as a function of our minds. This is the first step beyond a metaphysics of time and toward a phenomenology of temporality. Kant is, of course, a metaphysician and he does want to say that there are respects in which we must view ourselves as standing outside of time. In the moral sphere, for instance, when we judge an action to be right or wrong, we do so by projecting a conception of ourselves as moral legislators who are above time, deciding forever and always on the moral rule involved in action. In the metaphysical sphere, furthermore, Kant does argue for the existence of an immortal soul, although from a moral point of view only. Although we cannot have knowledge of our immortality, Kant maintains that we have to believe that we have an immortal soul insofar as we believe we can be moral agents. The argument starts from the premise that we cannot try to do something we believe to be impossible. Insofar as we act morally, we are trying to achieve something roughly like moral perfection. But because moral back-sliding is always possible, achieving this end would require an infinite amount of time. Therefore, wanting to be moral requires us to believe that we have immortal souls.

These considerations are pertinent to the present inquiry, however, only to the extent that they indicate some reasons Kant may have for saying that the self is both constrained by time and independent of time. If the mind is the origin of time, that does not make time any less real *for us*. The finitude of the mind is characterized not by the limitations on life, but by the time-bound nature of experience. Time is an a priori condition of every experience, even if it is not thematized in the experience.

What does this account of time tell us about Kant's understanding of temporality? For one thing, turning idealism's game against itself shows that whereas Cartesianism holds that "the only immediate experience is inner experience" and that outer experience is only mediated or inferred, for Kant the only immediate experience is outer experience and inner experience is only mediated (B276–277). Kant does not think that this turning of the tables means that we are not conscious of our own existence. That minimal sense of subjectivity is still preserved. Only a very minimal sense is preserved, it must be noted, because all that follows is *that* a subject exists. We are told nothing about *what* it is. That is to say, we do not thereby have experience (empirical cognition) of anything about the subject in itself (B277). All this reversal of Cartesianism entails is that "inner intuition, i.e., time" is possible only because outer objects are known to exist immediately (B277). Kant also maintains that persisting *matter* is not inferred a posteriori or "drawn from outer experience" (B278). On the contrary, it is an a priori presupposition as "the necessary condition of all time-determination, thus also as the determination of inner sense in regard to our own existence through the existence of outer things" (B278). "Persistence" is explained a priori (as substance), and is not obtained from outer experience. Persistence is not actually perceived, but it is a condition for the possibility of any particular perception (e.g., perception of change).

In the "Refutation of Idealism" the crucial question concerns why Kant thought that the persistent had to be external substance. Why could the persistent not be something more "inner" rather than "outer," more "subjective" than "objective" (to use some problematic terms)? Two perfectly good internal candidates for the persistent (or the "permanent" according to some translations) in experience are time and the "I think." Let me discuss time first. According to the "Transcendental Aesthetic," time is the essential feature of inner sense, and all experience involves inner sense (whereas only some experiences involve outer sense). Time is therefore a feature of every experience. Would not time be, then, a good candidate for the permanent backdrop for perception, which is, of course, not perceived as such? Kant's rejection of this possibility is stated forcefully and clearly. "Time," he writes in the A edition, "has in it nothing abiding, and hence gives cognition only of a change of determinations, but not of the determinable object" (A381). Here there might seem to be a metaphysical issue about the nature of time insofar as this claim that time "has in it nothing abiding" seems to contradict his other claim that time is the permanent framework that makes experience possible. Note, however, that he says "*in it*" (where "it" refers to time). Does that tell us whether time itself is changing or unchanging? One current reading is that the framework of time is always there (although it is not perceived), but within that framework the content is always changing. His argument is drawing not so much on the metaphysics of time per se, however, as on the phenomenology of time-determination. If time changed every moment, then there would be nothing that could feature in each and every experience. The point is rather that time, which has in it "nothing abiding," could not be determined, that is, experiences could not be ordered, except against an unchanging backdrop, which must be substance and not time.

Accepting this argument does not lead right away to the confirmation of external substance as the permanent backdrop. Another internal candidate could be the "I think" itself. In fact, insofar as the "I think" must be able to accompany all my experiences, and is thus a permanent framework for experience, it would seem to be an even better candidate for the permanent. Kant rejects the cogito as the source of persistence, however, for much the same reason as he rejects time as the permanent. The above quotation then continues, "For in that which we call the soul, everything is in continual flux, and it has *nothing abiding*, except perhaps (if one insists) the I, which is simple only because this representation has no content, and hence, no manifold, on account of which it seems to represent a simple object, or better put, it seems to designate one" (A381; emphasis added). There is "nothing abiding," then, either in time or in the mind. In the Paralogisms he also asserts: "But now we have in inner intuition nothing at all that persists, for the I is only the consciousness of my thinking" (B413). In context, his reason for asserting that permanence is not given in inner intuition is that he wants to show that the oneness or unity of consciousness does not prove the existence of a permanent self (B420). The unity is not an intuition of the subject as object (B422). The purely formal "I" is the same in every experience, and does not have any content that could stay the same. Persisting or abiding content is required if I am to be able to perceive temporal difference, for instance, by determining that there were two separate events and that one came before the other. Thus, he says that the representation "I" is not an intuition but "a merely *intellectual* representation of the self-activity of a thinking subject" (B278). As such an empty thought, the "I" provides nothing that could be the basis for the persistence that makes possible the perception of motion and change.

Insight into Kant's understanding of the nature of subjectivity can be gained most directly from the Transcendental Deduction of the first *Critique*. What is "deduced" in that section of the Transcendental Analytic? In contrast to the Refutation of Idealism, which shows that there is no "I" without an "It," the Transcendental Deduction can be summed up as a proof that there is no "It" without an "I." These slogans may be useful pedagogically to sum up Kant's complex and prolix text, but they can also be misleadingly simple. For instance, the "I" in each case is different. The "I" in the Refutation of Idealism is the subjectivity that can be introspected, the empirical ego. In contrast, the "I" in the Transcendental Deduction is the transcendental ego, the subjectivity that is doing the introspection. This difference could also be characterized as the difference between the constituting consciousness and the constituted consciousness. Kant wants to establish that whereas the input through sensible intuition is a manifold, the output that is actually experienced (whether inner or outer) has a unity to it (or better, a oneness). Where does the oneness come from? It could not come from the intuitions, which are a multiplicity. Even the concepts are multiple. The oneness of experienced output, on this model, would not be possible unless a *single* processor synthesized the manifold.

Clearly this metaphor of a combinatory processor has its limitations, however helpful it might be in revealing the differences between Kant and his predecessors. There are questions, for instance, about whether the hardware or the software is the source of the oneness or unity. Even if the software is the processor, there is still a question about whether the metaphor captures distinctions about consciousness adequately. The relation of the transcendental ego and the empirical ego, for instance, is not to be thought of as the relation of a container to the contained. Admittedly, it is hard not to think of the relation that way when Kant himself says things like "all manifold of intuition has a necessary relation to the 'I think' in the same subject *in which* this manifold is to be encountered" (B132). This way of putting the point makes the introspected content seem as if it is encountered "in" the mind. In the same breath Kant will also reverse the containment relation and make it seem as if the "I think" is contained in experience, when he says that "in all consciousness [the 'I think'] is one and the same" (B132). The little word "in" is thus troublesome insofar as it can suggest the relation of spatial containment, which Kant does not want to imply, as well as what he does mean to suggest, which is more on the order of logical implication. Perhaps this line could have been better rendered in English as, "*throughout* all consciousness the 'I think' is one and the same." The German does say, however, *"in allem Bewusstsein ein und dasselbe ist,"* so both the Kemp Smith and the Guyer/Wood translations are correct to use the word "in." A careful reader should be aware, however, that the word does not necessarily connote spatial containment.

   In sum, the principle of persistence is not and cannot be "in me," and it cannot be either "the I of apperception" or "time." Inner sense is constantly changing, but to be able to say this, there has to be something that is not changing (B277). This cannot be the "I," because there is no intuition of the "I." Kant concludes then that the "I think" as a persisting formal framework would be empty of content and would not suffice as the persistent background for temporal discrimination. How would one know, say, that there were two empty moments in succession? Nothing plus nothing is nothing.

Heidegger's Reading of Kant

Where, then, does time come from? What connects the stream of consciousness? What makes experience a unity such that we can know that time is continuous and that there is only one world? Kant's masterful move is to claim to be *both* a realist and an idealist, but not in the same way. Here is where the previously mentioned distinction between empirical realism and transcendental idealism comes in. Kant wants to be an empirical realist, and thus neither a dogmatic empirical idealist like Berkeley, who denies external substance, nor a problematic empirical idealist like Descartes, who doubts the external world. Kant maintains that the only way to be an empirical realist is to be a transcendental idealist.

What does it mean to be a transcendental idealist specifically about the nature of time? Perhaps the most radical answer to this question in the history of the reception of Kant is Martin Heidegger's reading of Kant on time in *Kant and the Problem of Metaphysics.* Published initially in 1929, shortly after his publication in 1927 of his major work, *Being and Time*, this so-called *Kantbuch* is intended to provide a more basic understanding of philosophy by revealing the links between Kantian transcendental philosophy and Husserlian phenomenology. These were the dominant approaches to philosophy in Heidegger's day, and what they both missed, according to Heidegger, was the fundamental importance of temporality. Heidegger was intensely preoccupied with time during the 1920s. Kant's writings on time provided the crucial backdrop for Heidegger until he foregrounded them in this study. In particular, Heidegger's Kant book can be considered as a study of the section of the first *Critique* called the "Schematism." This section of the *Critique* explains how time is added to intuitions and concepts as the transcendental machinery cranks out experience through the various levels of processing. In any case, Kant clearly had become the test for any philosophical account of time.

In this section I will argue two theses. First, I will try to explain briefly why even if the Kant book is mistaken as a reading of Kant, it nevertheless illustrates the difference between the Husserlian and the Kantian approaches to time constitution. Second, I want to establish that this reading of Kant shows how Heidegger's project of explaining temporality "errs" as a general philosophical project. "To err" is not the same as to be in error in a way that could lead, say, to failing a test. It could also mean something like going in a different direction from standard ways of thinking, or uncovering insights that are buried in the text. In this section, I will be turning the charge of errancy back against not only Heidegger's reading of Kant, but also his attempt to make temporality the foundation of metaphysics. Heidegger acknowledged the first mistake. Whether he ever saw the second errancy is more difficult to determine.

Heidegger's interpretation of Kant was intended to be a later part of *Being and Time*, of which he published only a part. In the author's preface to the first edition of *Kant and the Problem of Metaphysics*, Heidegger says, "This interpretation of the *Critique of Pure Reason* arose in connection with a first working-out of Part Two of *Being and Time*."[6] In this part he intended to deconstruct and even to destroy the history of philosophy through a series of readings that would show where previous philosophers failed to do philosophy right, that is, where they fell short of doing philosophy in Heidegger's way and therefore where they went wrong in their analyses of fundamental phenomena, particularly time and temporality. In *Being and Time* he says that the purpose of this destruction is not simply to "shake off" the tradition, but to shake it up. "This hardened tradition must be loosened up," he says, in order to discover new possibilities that are contained in it but that have been occluded by the standard interpretations.[7]

In the author's preface to the second edition of the Kant book in 1950, Heidegger acknowledges the charge of Ernst Cassirer and other critics that his readings do violence to the historical texts. He justifies this violence as the supposedly inevitable result of trying to engage the texts in a thinking that could give rise to new philosophical insights. "A thoughtful dialogue," he remarks, "is bound by other laws."[8] The "other laws" are presumably the laws not of accurate philology but of good philosophy. He then gives his *mea culpa*: "The instances in which I have gone astray and the shortcomings of the present endeavor have become so clear to me on the path of thinking during the period [since its first publication] that I therefore refuse to make this work into a patchwork by compensating with supplements, appendices, and postscripts. Thinkers learn from their shortcomings to be more persevering."[9]

At issue in Heidegger's reading of Kant is the importance that Heidegger gives to the faculty of the mind Kant calls the imagination. Kant himself deemphasizes the role of the imagination in the B edition of the *Critique of Pure Reason*. Heidegger faces a long tradition of Kant scholarship that maintains that the B edition transforms the "psychological" arguments of the A edition into more properly "logical" arguments. Heidegger, in contrast, sees the B edition as even more psychological than the A edition, and in any case, for Heidegger the distinction between the psychological and the logical misses the point of both editions, which is to be "transcendental."[10] The transcendental is both "subjective" and "objective," depending on whether the focus is on inner or outer experience. On Heidegger's reading, the main difference between the two editions is the shift from the pure power of the imagination to the pure understanding as the central faculty of "transcendence." Transcendence is synonymous with the "possibility of experience."[11] In *The Metaphysical Foundations of Logic*, Heidegger explains that transcendence, or Being-in-the-world, in contrast to intentionality for Husserl, is not a movement from interior to exterior.[12] Transcendence first constitutes the subjectivity of a subject and makes the intentional distinction between interior and exterior possible.

Kant scholars must of course heed the self-understanding of the master, and thus Cassirer and others cannot take Heidegger's reading seriously. From the point of view of a history of ways of understanding temporality, however, Heidegger's Kant book represents a unique account of time constitution. Whether the theory advanced is Kant's own understanding or merely Heidegger's "errancy" is beside the point. For present purposes, Heidegger's Kant book can be regarded as confronting two different traditions of theorizing the connection between time and the mind. The first approach is through faculty psychology. This is the Kantian approach and it is based on an understanding of the mind as the interaction of what Kant called "faculties." Faculty psychology sees time as being added by one particular part of the mind to the output of each and every moment of experience.

In contrast to this atomistic account of the source of time, there is a more holistic model of the mind that sees time differently.[13] Call this the "duration" account because it accounts for time as duration rather than as a series of moments. The two principal theorists of duration that I will be discussing in more detail below are Husserl and Bergson.

Let us look first at Kantian faculty psychology. The Kantian approach of faculty psychology sees different faculties as having different functions. Sensibility, for instance, contributes the data brought under concepts by the faculty of the understanding in what Kant calls synthesis. Whereas the understanding always involves synthesis with sensibility, the faculty of reason applies concepts independently of sensibility. In addition to these three faculties, Kant also sometimes speaks of a fourth faculty, the imagination. In the first edition the imagination plays a more central role than it does in the second edition, where it is no longer described as a separate faculty (although its importance is reestablished in Kant's discussion of judgment, including aesthetic judgment in particular).

In the A edition of the first *Critique* Kant maintains, according to Heidegger, that what orders experience temporally is not sensibility, understanding, or reason, but the imagination. The imagination adds time to the synthesis of intuitions (data) and the categories (concepts). As a result of such a synthesis, each moment of experience is a unit of a single time.

In today's terms the Kantian faculties might be called "modules." Using metaphors for the mind drawn from computer science, contemporary cognitive psychology often speaks of modules that operate below the level of consciousness. These modular "subprocessors" then filter the data and "synthesize" or process it into a form recognizable by a higher-level processor. Kant hypothesizes three levels of such cognitive processing or synthesis. Each one of these levels of synthesis is, in Heidegger's terms, "time-forming." Heidegger's claim is that these activities are the source of time in its various dimensions. Although he cannot find much textual evidence in the first *Critique* itself, he finds at least some grounds in Kant's *Lectures on Metaphysics* for thinking of these three syntheses as forming respectively the three temporal modalities of present, past, and future.[14]

Thus, the first level is the synthesis of apprehension. This is where the data get entered. Perception is a paradigm case of this type of synthesis. More to the point, this level of synthesis produces or forms time as the series of Nows. It is thus the source of the present with which we "reckon," even if "this sequence of nows, however, is in no way time in its originality."[15] By the term "original," I take Heidegger to be saying that for Kant the *source* of time is the transcendental power of imagination, which allows this experience of time as a sequence of Nows to "spring forth."[16] Heidegger underscores the role of imagination in the formation of time when he says, "If the transcendental power of imagination, as the pure, forming faculty, in itself forms time—i.e., allows time to spring forth—then we cannot avoid the thesis [that] the transcendental power of imagination is original time."[17]

The synthesis of reproduction takes place when the input is reprocessed in the absence of the source of the data. Reproduction is "bringing-forth-again," and is, accordingly, "a kind of unifying."[18] What kind of unifying does Heidegger mean, exactly? The argument he gives is that the mind must not "lose from thought" that which "differentiates time."[19] In other words, if the mind did not know the difference between thoughts that it was having now and thoughts that it had earlier, that earlier experience would be lost completely. Heidegger sees this mode of synthesis as essential to the oneness and unity of experience:

the pure power of imagination, with regard to this mode of synthesis, is time-forming. It can be called pure "reproduction" not because it attends to a being which is gone nor because it attends to it as something experienced earlier. Rather, . . . it opens up in general the horizon of the possible attending-to, the having-been-ness, and so it "forms" this "after" as such.[20]

Whereas the synthesis of apprehension forms experience into a sequence of Nows, the synthesis of reproduction adds the possibility of forming time into past as well as present times. The question then arises, is this characterization of time sufficient, or is a third form of synthesis needed, one that forms time into the future? Will this formation of time be as essential to experience as the present and past are?

Heidegger would like the text to show that the future is formed in the synthesis of recognition, which is the level where self-consciousness begins to play more of a role. He admits, however, that there is little or no textual evidence in the first *Critique* for the temporal interpretation that he wants to give the synthesis of recognition as futural. Indeed, Anglophone commentators often read the synthesis of recognition as an argument for the necessity of the transcendental unity of apperception, which is in some sense outside of or independent from time. Heidegger's preoccupation with time leads him to read Kant's argument for the synthesis as amounting to an argument for the need for the future in order to make sense of the analysis that was just provided for the syntheses of apprehension and reproduction. Heidegger therefore claims that although the synthesis of recognition is third in the order of exposition of the syntheses, in terms of logical priority it comes before the other two syntheses. Heidegger sees the third synthesis as "in fact the first."[21] "It pops up in advance of them," he asserts, and the arguments for the necessity of *Abbildung*, or likeness, and *Nachbildung*, or reproduction, depend on the argument for *Vorbildung*, or prefiguration.[22] Let's see how Heidegger forces a temporal dimension on Kant's text.

"Without consciousness that that which we think is the very same as what we thought a moment before," writes Kant, "all reproduction in the series of representations would be in vain" (A103).[23] Heidegger adds that something could not be thought to be the same except against a backdrop that also remains the same. This empirical claim leads to the idea of a more general or "pure" horizon of "being-able-to-hold-something-before-us [*Vorhaltbarkeit*]."[24] This *Vorhaltbarkeit* amounts to a *Vorhaften*, a preliminary attaching or a prefigurative grasping. The "*vor*" suggests a projection of a future in this fore-structuring of experience. Heidegger therefore concludes that the synthesis of recognition is time-forming and the time that it forms is the future: this synthesis, he says, "explores in advance . . . what must be held before us in advance as the same in order that the apprehending and reproducing syntheses in general can find a closed, circumscribed field of beings within which they can attach to what they bring forth and encounter, so to speak, and take them in stride as beings."[25] Because the first two syntheses presuppose this third synthesis, Heidegger believes that he can even maintain that the future has logical priority over the present and past. He thus derives from Kant a transcendental argument for the primacy of the future. The argument is that because there is no self without time, and no time without a future, therefore, there is no self without a future. This argument is remarkably different in character from the argument for the primacy of the future that he developed in *Being and Time*. There he showed the priority of the future through the more existential account of being-toward-death. Discussion of these different approaches will have to wait until chapter 4, which deals with the future.

For now, I need to explain the case that Heidegger makes for the first premise, which concerns the relation of time to the self. Given the ideas of time and the "I think," which is the source of which for Kant? Heidegger discusses this issue in reference to Kant's famous sentence about the mind-dependency of time: "Time is therefore merely a subjective condition of our (human) intuition (which is always sensible, i.e., insofar as we are affected by objects), and in itself, outside the subject, is nothing" (A35/B51). If Kant appears to be pulling the rug out from underneath himself here, one must remember that in addition to being an empirical realist about time, he is also a transcendental idealist, and it is as the latter that he is speaking at this point in the text.

Taking off from this striking claim, Heidegger provides an even more astonishing account of time as the source of the self. As a faculty psychologist, Kant is normally thought to be saying that time is subjective in the sense that the subject generates experience by imposing the form of time on the data of intuition. Heidegger, however, reverses the relation and suggests that time is the source of subjectivity. He makes a good point when he says that time is not something that affects a self that is already "at hand." The self is not a distinct object or, as Heidegger would say, a *vorhanden* present-to-hand thing, to which time could then be attributed as if time were a property that an object could or could not possess. Heidegger then suggests, however, that "time as pure self-affection forms the essential structure of subjectivity."[26] A thoughtful reader might well wonder whether time is the sort of thing that could be a self-affecting activity or that could turn into subjectivity. But on Heidegger's reading, one thing should be clear, namely, that Kant is neither an idealist nor a realist about time. The debate between realists and idealists is about whether the mind or the world is the source of time. An idealist maintains that time is imposed by the mind on experience. An idealist could not make the curious assertion that Heidegger attributes to Kant, namely, that time is what makes self-consciousness possible,[27] and that it "first makes the mind into a mind."[28]

If an idealist could not make this assertion that time is the source of mind, could a realist make it? One might think so, because if time comes before subjectivity, then it is more real than subjectivity. Insofar as realism says that to be real is to be in time and space, however, this position could not be a form of realism. To say that time was real would be a category mistake that confused a necessary condition of reality with something that was itself real. In any case, the idea that temporality is the source of time raises questions that are prior to the realist–idealist debate.

Heidegger therefore positions his reading of Kant before the distinction between realism and idealism can get a foothold. Time is not *in* the mind, but rather is the ground for the possibility of the mind and the self. Because the temporal movement " 'from-out-of-itself-toward . . . and back-to-itself' first constitutes the mental character of the mind as a finite self," time and the "I think" are not at odds with each other, but "they are the same."[29] What does this mean? One thing to note is that when Heidegger speaks of "the mind," he is speaking loosely, insofar as his theory does not allow him to use the term, and it is not a technical term of Kant's either. Another point to note is that Heidegger is not identifying the transcendental unity of apperception with "the mind." For him, the mind is empirical consciousness, whereas the "I think" is not a content of consciousness but rather a condition of it. In the previous quotation Heidegger even says explicitly that the pure self-affection of original time is not the self-positing of a preexisting mind among others, but rather that it "first constitutes the mental character of the mind as a finite self."[30] Thus, subjectivity does not exist prior to original time, but is made possible only through original time. Both time and the I of pure apperception are said to be fixed, unchanging, and perduring.[31] These characteristics are usually attributed to mental substance, but Heidegger's Kant does not believe in mental substance. Heidegger is instead hypothesizing that what Kant really wants to say is the following:

for Kant only wants to say with this that neither the I nor time is "in time." To be sure. But does it follow from this that the I is not temporal, or does it come about directly that the I is so "temporal" that it is time itself, and that only as time itself, according to its ownmost essence, does it become possible?[32]

Heidegger grants that this interpretation does violence to Kant,[33] that Kant does not expressly see this himself,[34] and that Kant was "unable to say more about this."[35] Heidegger then points to his own *Being and Time* as the standpoint from which to see how laying the ground for metaphysics "grows upon the ground of time."[36] Heidegger's turn away from the Kantian style of philosophy and especially from the use of theoretically laden terms such as "subjectivity," "consciousness," and even "experience" is motivated by an increasing skepticism about the idea of experience experiencing itself. The point of *Being and Time* is to avoid the Cartesian problems that result from using these terms, and to create a new vocabulary for phenomenological analysis. This change of vocabulary will enable Heidegger to think about issues of time and temporality differently, both in style and in substance. Now is the time, then, to turn to Heidegger's own phenomenology of temporality in *Being and Time*, with some considerations about the development of his innovative theory.

The Early Heidegger

Although Heidegger began publishing on time as early as 1915 in "The Concept of Time in the Science of History," his analyses more clearly resemble those of *Being and Time* in lectures from 1924 and 1925, including *The Concept of Time*, *The History of the Concept of Time*, and also "Wilhelm Dilthey's Research and the Struggle for a Historical Worldview." There are also important clarifications in lectures given shortly after *Being and Time*, including *The Basic Problems of Phenomenology* (1927) and *The Fundamental Concepts of Metaphysics: World, Finitude, Solitude* (1929). The foregoing explication of Heidegger's analysis of Kant in *Kant and the Problem of Metaphysics* left us hanging on Heidegger's curious but fascinating remark about time and the I of apperception being the same. To see what he means and why he said what he did, it is important to understand Heidegger's phenomenology of temporality, especially in *Being and Time* and these other early writings on temporality. Heidegger's intention is to show that Kant's way of thinking about time and space is derived from what Heidegger calls a more "primordial" level of questioning. In contrast to Kant's transcendental arguments, which show that if something is required for knowledge then something else is also required, Heidegger's derivation arguments try to reverse the ontological ordering of the terms of analysis. From the Kantian perspective, time in the objective Newtonian sense of the present-to-hand (*vorhanden*) universe comes before (i.e., is logically prior to) the human, qualitative experience of temporal moments. Heidegger inverts that ordering and argues not that objective, clock time does not exist, but that objective time is not intelligible without Dasein's prior qualitative temporality. Heidegger's project in *Being and Time* (1927) is to show that starting from objective time, the philosopher will not be able to explain qualitative temporality, but starting from qualitative temporality, the philosopher can explain objective time.

In the Dilthey paper of 1925 Heidegger remarks that "we ourselves are time."[37] Because at that point Heidegger does not distinguish consistently between "time" and "temporality," there is an ambiguity in this claim that we are time. From this assertion what is not clear is whether it is the public "we" or each private individual that is time. When he says in his more technical language, therefore, that "in each case Dasein itself is time," the phrase "in each case" suggests that time is relative to each particular Dasein.[38] This clarification leads to a further problem, however, insofar as it implies that there are as many different times as there are lives. This claim would be hard to reconcile with the standard Kantian intuition that time is one.

To sort out this problem, we must first ask whether the "is" in the expression "Dasein is time" is the "is" of identity or the "is" of predication. Heidegger should not mean the "is" of predication, or he would be back in the Kantian camp of faculty psychology whereby time is a feature that is applied by one faculty (whether the imagination or the understanding) to another faculty (sensibility). That Heidegger means something as strong as an identity claim is indicated when he says, "Human life does not happen in time but rather is time itself."[39] In his more technical language he writes, "The *being-there* of Dasein *is* nothing other than *being-time*. Time is not something that I encounter out there in the world, but is what I myself am."[40] Thus, time is encountered neither as an entity outside in the world, nor as something that whirs away inside consciousness. On this formulation I note that it also does not seem possible to ask which comes first, Dasein or temporality. As a result, the neo-Kantian effort in *Being and Time* to "deduce" one from the other turns out to be unnecessary.[41]

Nevertheless, Heidegger offers a reasonably straightforward argument for the prioritization of temporality over Dasein. *Being and Time* states clearly that "Time is primordial as the temporalizing of temporality, and as such it makes possible the Constitution of the structure of care."[42] "Care" is a technical term that means that Dasein is always a being-in-the-world whose relation to the world makes Dasein what it is. In other words, Dasein is necessarily care. Heidegger's first premise is thus that time makes care possible.[43] He then infers from the fact that care is what Dasein is that time also makes Dasein possible. Heidegger maintains further that temporality's temporalizing makes possible "the multiplicity of Dasein's modes of Being, and especially the basic possibility of authentic or inauthentic existence."[44] Temporality thus leads to making the distinction between authentic and inauthentic, as an example will soon illustrate. Although Heidegger denies that "authentic" and "inauthentic" are value-laden terms, they clearly indicate different ways of caring. Authenticity is a way of caring about death, whereas inauthenticity tries not to care about it. These different ways of caring could be called "normativity," and thus temporality is shown to make normativity possible.

One problem with this argument is that if Dasein is care, then by saying that care is possible only through Dasein's temporalizing, Heidegger seems to be caught in a tautology. He would then be saying vacuously that Dasein makes Dasein possible. Heidegger's attempt to work out this puzzle is advanced somewhat by his analysis in *The Basic Problems of Phenomenology* (1927), which are lectures that he gave during the year in which *Being and Time* was published. *Basic Problems* distinguishes between *Temporalität* and *Zeitlichkeit*. Both are translated as "temporality," but Albert Hofstadter, the translator, capitalizes Temporality when it means *Temporalität* and lets the lower-case stand for *Zeitlichkeit*. Thus, *Temporalität* is the Temporality that makes a priori knowledge of the objective possible and *Zeitlichkeit* is the ontological temporality of the understanding of being. The lectures break off before this distinction can be developed much more than to say that "time" is the most a priori phenomenon, "*earlier than any possible earlier* of whatever sort, because it is the basic condition for an earlier as such."[45]

This argument is problematic on two counts. First, in using the term "time" here, Heidegger's claim becomes ambiguous because it does not specify which of the two senses of time is meant. One assumes that by "time" in this sentence he means temporality in the sense of *Zeitlichkeit* insofar as this is what temporalizes itself (*sich zeitigt*). Second, Heidegger maintains that the term "a priori" means "earlier" in a temporal sense. "A priori" means "earlier," he says, and "earlier" is "patently a *time-determination.*"[46] This claim could well be suspected of confusing the "temporally prior" with the "logically prior." He explains, however, that he does not mean to say that a priori conditions are "temporally" prior in the sense where "temporally" implies "before" in the ordinary, "intratemporal" understanding of time as a succession of moments in which we stand.[47] But at the more fundamental level where "temporality [*Zeitlichkeit*] temporalizes itself,"[48] he mocks the tradition for not realizing that "it cannot be denied that a time-determination is present in the concept of the a priori, the earlier."[49] Even with this qualification, though, Heidegger still appears to be confusing "priority" in the logical sense with "priority" in the temporal sense. Although Heidegger is thus wrong, given current practice, on this question of word usage, he is right on the more important point that the aprioricity of the Temporal does not make it ontically the first being, because time is not a being at all. As such, time cannot be said to be ontically "forever and eternal."[50] What is normally thought to be the case when ontic time no longer obtains is not clear in any case. Cold ashes in the motionless void, one supposes.

In *The Fundamental Concepts of Metaphysics* (1929), Heidegger adds some analyses that illustrate and indirectly clarify his statement that "*temporality temporalizes.*" The topic is boredom. This starting point might seem to be an arbitrary and inauspicious basis for a theory of human existence insofar as boredom is merely one among many subjective states of mind in which one can find oneself. Heidegger's intention, however, is to show how attending directly to a phenomenon like boredom and avoiding the Cartesian vocabulary of consciousness will be more useful than assuming from the start that boredom is a merely subjective state of mind. Such an assumption presupposes an unbridgeable gap between subjective experience, to which the subject is the only one who has access, and objective experience, which is accessible from many points of view, including natural science.[51]

Heidegger challenges the method that presupposes this gap. He maintains that this method treats our access to consciousness as itself something that can be made into an object for what he calls an "ascertaining" consciousness. "Ascertaining" tries to bring consciousness itself to consciousness and does not recognize that this objectifying attempt in fact alters or destroys the phenomenon in question.[52] Instead of this mode of false reification, Heidegger argues for a phenomenological approach that he calls "awakening." The ascertaining consciousness depends on a more basic implicit understanding to which we can be awakened. In awakening, the phenomenon in question is described not to objectify it and bring it under our control, but to *release* it from the grip of Cartesian and Kantian theories based on the notions of subjectivity and consciousness.

Theories of consciousness focus primarily on cognition, and they tend to treat other phenomena such as moods or emotions as side issues. Heidegger, in contrast, attributes greater importance to moods and emotions, which are a function of the basic category or "existentiale" of Dasein that he calls *Befindlichkeit* (disposedness). Our *Befindlichkeit* is a function of how we find ourselves in the world, how disposed or attuned we are to the situation that enables us to be who we have been. In *The Fundamental Concepts of Metaphysics* Heidegger specifies disposedness as a *Grundstimmung*, which means that we are never without some emotive attunement. Because it is neither entirely conscious nor unconscious, however, this basic attunement needs to be "awakened" rather than "ascertained." The attempt to ascertain attunement by making it explicit only serves to diminish it.[53] Not an experience in the "soul," attunement reflects how we are there—Da—in the world with one another.[54] In contrast, *Verstehen* or the Understanding involves projecting possibilities into the future as the basis for action. Heidegger does not assume from the start that

attunement is a merely subjective phenomenon, unlike most philosophers who see moods as only subjective. Neither merely subjective nor entirely objective,[55] modes of attunement reveal how
we find ourselves in a particular situation that both conditions what
we can do and delimits what cannot be done. In *Being and Time*
Heidegger focuses his discussion of attunement on fear and anxiety.
Here in these lectures, *The Fundamental Concepts of Metaphysics*,
he works instead on the mood of boredom. His intention is to show
the particular connection between mood and the experience of
time.

The German word for boredom is *Langeweile*, or literally a
"long while." Using boredom as the paradigm instead of anxiety,
Heidegger argues that time is lengthened by boredom, and he
describes some of our strategies for evading boredom by "shorten-
ing" time. With this etymological analysis of the German word for
boredom, however, Heidegger runs the risk of an overly psychological argument for his derivation claim. Just because the human experience of time can be long or short, it does not follow that human
temporality is more *primordial* than objective time. At this point,
though, it becomes important to ask, what does "primordial" mean
for Heidegger? The term can be used in at least two ways. In one
sense, it means "most basic" or "ground." In another sense, however,
it merely means "without which." The former implies that if something, call it (a), is more primordial than something else, call it (b),
then (a) could obtain when (b) did not obtain. The second usage is
weaker and says only that there could not be (b) unless there were
(a), but not that (a) could obtain even if (b) did not. On my reading,
Heidegger holds the weaker relation between time and temporality:
we could not reckon with objective time without existential temporality, but temporality is not so basic a level of experience that
temporality could obtain in the absence of objective time.

In *Fundamental Concepts of Metaphysics* he acknowledges that
attunement concerns specific individuals, and thus seems psychological. He also thinks, however, that as "primordial," profound

boredom makes psychology and psychoanalysis both possible and necessary. This boredom permeates "modern man" generally and is the mood or attunement of the present age. Heidegger has thus generalized Kierkegaard's account and extended it into a critique of modern subjectivity.

Profound boredom contrasts with two other forms of boredom. Each of these three forms of boredom has two structural features by which it manifests its concern for things and its care for itself: (1) being held in limbo, and (2) being left empty. These can also be viewed as strategies for relating to boredom. The first form of boredom is "becoming bored *by*" something in the world. This form of boredom seems to be caused by an object "outside" oneself such that we complain, "It's boring!" Heidegger's own example is a long wait for a train. One tries to escape this form of boredom by "passing the time." For instance, while waiting for the train one might find oneself constantly looking at one's watch. Another example might be a philosophy lecture on boredom. As the lecture drags on, one might find oneself watching how slowly the second hand of the clock on the wall moves around the dial—as if this activity will "shorten" time and make it go by faster. The characteristics of this form of boredom are the *wearisome* and the *tedious.* One is held in limbo by the wearisome situation, and the tediousness of the things that refuse to conform to one's wishes and expectations leaves one empty.

If the world is the source of this first form of boredom, the source of the second form of boredom is more explicitly the self. This form of boredom is "being bored *with*" one's self in its situation. Heidegger's example is of a social evening that seemed to be a pleasant experience while it was occurring, but later one realizes how bored one was. Here the time is passed differently insofar as there is no behavior such as constantly looking at the clock to make time pass more quickly. Whereas the first kind of boredom arises from the world, the second form arises from Dasein.[56] We do not say, for instance, as we might have in the first form of boredom, "the book was boring." Instead, now we would speak of ourselves as being boring to ourselves. Heidegger does not say so, but the quality of being boring to oneself could well make one boring to others. One sees that there is no reason in the world for being bored with the social evening, but nevertheless, one is. One is held in limbo by the standing of time, or what I will call "taking one's time," as if time were a commodity that one could dispense at will. By taking one's time and wasting it as one wants, one thereby hopes to bring time to a standstill and to halt its flow. That is to say, we try to forget the future and the past, and we try to convince ourselves that all that counts is the present. But in fact time is again not under our control, and it does not vanish.

Heidegger's third form of boredom is "profound boredom." In this form we are indifferent both to the world and ourselves. Moreover, the connection between the self and temporality becomes markedly evident. Being held in limbo occurs in this case through the refusal to come under one's control not of some particular thing, but of things *as a whole*.[57] This way of being held in limbo leads to our being left empty in the form of a bemusement with time as a whole. The emptiness is a function of the withdrawal of everything, and the inability of anything to engage our interest and involvement. Playing on the idea of sightings (*Sichten*), Heidegger describes how this withdrawal takes place in each temporal dimension:

All beings withdraw from us without exception in every respect [*Hinsicht*], everything we look at and the way in which we look at it; everything in retrospect [*Rücksicht*], all beings that we look back upon as having been and having become and as past, and the way we look back at them; all beings in every prospect [*Absicht*], everything we look at prospectively as futural, and the way we have thus regarded them prospectively. Everything—in every respect, in retrospect and prospect, beings simultaneously withdraw. The *three perspectives* [*Sichten*] of respect, retrospect, and prospect do not belong to mere perception, nor even to theoretical or some other contemplative apprehending, but are the perspectives of all *doing and activity* of Dasein.[58]

Through this withdrawing, we gain for the first time a standpoint on the entirety of all that is withdrawing. Everything has to start to withdraw for us to get a sense of the whole. Heidegger calls this grasp of the whole the *Augenblick* or moment of vision in which the unity of one's temporality is grasped as an integral existential possibility.

Today one might think that the phenomenon that Heidegger describes as profound boredom is really clinical depression. If that were so, there would be a fairly straightforward way out of this all-pervading boredom, namely, to take an antidepressant. Heidegger would regard this manner of responding as a failure to appreciate the way attunement reveals the world as such to Dasein. To see Heidegger's answer in 1929 one must distinguish the concept of Dasein from that of both self and subjectivity. What he wants to do is *"not to describe the consciousness of man but to evoke the Dasein in man."*[59] What is the difference between describing consciousness and evoking Dasein? Heidegger apparently believes that the realization that "modern man" is fundamentally bored with existence causes people to thematize the difference between their existence and the concept of "man" as subjectivity. Boredom is precisely the gap between subject and world that makes "man" possible. Realizing this "fundamental attunement" of the present age leads us, he says, to want "to liberate the humanity in man, to liberate the humanity of man, i.e., the *essence* of man, *to let the Dasein in him become essential.*"[60] He then elaborates on this demand for liberation as follows: "This demand has nothing to do with some human ideal in one or other domain of possible action. It is the *liberation of the Dasein in man* that is at issue here. At the same time this liberation is the task laid upon us to assume once more our very Dasein as an actual *burden.*"[61] Dasein must learn to answer for itself, and philosophy plays a crucial role here by getting the Dasein to realize that it has to take on the burden of being free. Only if Dasein takes this burden on itself will it be able to do something concrete about its situation.

Even the philosopher has a mood, of course. But Heidegger attributes mood not simply to the philosopher, that is, the person doing the philosophizing, but to the philosophizing itself. People can be gripped by a fundamental attunement without recognizing it as such.[62] The world-weariness of ennui and *Weltschmerz* were standard conditions at the end of the nineteenth century, and indeed, ennui may even have been a nineteenth-century French invention. Nevertheless, the activity of philosophizing would not be revelatory if it were not itself grounded in a basic attunement. *"Philosophy,"* he emphasizes, *"in each case happens in a fundamental attunement."*[63] Citing Novalis, Heidegger suggests that modern philosophizing represents a fundamental attunement of homesickness. This mood of homesickness reflects philosophy's desire to be at home everywhere in the world, when it cannot be. Because people are not at home in the world at all any longer, the modern philosopher's mood is melancholic. For Heidegger there is no creativity without melancholy. That is not to say, however, that melancholy is always creative.

What is this depressing, profound boredom about? Heidegger suggests ironically that what is boring is neither objects nor subjects, even if these seem to be the only two possibilities. What is profoundly boring is time. More precisely, temporality, or the time of Dasein, is what is boring. In boredom, Heidegger says, "one feels timeless, one feels removed from the flow of time."[64] But this is an oppressive feeling. Boredom is oppressive because time weighs heavily. Profound boredom is ontological, and it makes ontic boredom possible. Ontic boredom is boredom with a particular thing or situation (for instance, being bored by a long discussion of boredom). At the same time, the occurrence of ontic boredom points to ontological boredom. Ontological or profound boredom is emptiness, where everything withdraws. This withdrawal of everything makes Dasein aware of the whole of its existence.

Boredom is thus as ontologically revelatory of the whole of our life as anxiety (Angst) is in *Being and Time*. Neither is about any particular thing, but each is about everything (and nothing). Neither of these, says Heidegger, is the *only way* to grasp the whole of one's existence. Heidegger points toward this conclusion in his 1929 lecture, "What Is Metaphysics?" There too he distinguishes between ordinary boredom and genuine or profound boredom. Of the latter he writes, "Profound boredom, drifting here and there in the abysses of our existence like a muffling fog, removes all things and human beings and oneself along with them into a remarkable indifference. This boredom reveals beings as a whole."[65] Joy and awe are said to offer a comparable revelation. Boredom, anxiety, joy, and awe each represent different ways in which our Dasein is revealed to us.

On my interpretation, though, anxiety and boredom have different effects on our self-understanding. Anxiety is said to *individuate* Dasein by making Dasein confront its unique fate and destiny. Individuation is an encounter with what is meaningful about the world, and it still involves what Heidegger calls existential care. Rather than *individuate* Dasein, however, I read Heidegger as suggesting that profound boredom *subjectivizes* Dasein. To say that Angst individuates is to say that each Dasein finds out what it cares about, and what makes it "in each case its own." To say that profound boredom subjectivizes is to maintain that the Dasein becomes indifferent to all meaningfulness and ceases to care about the world. Because of the degree of indifference to all meaningful interactions, the Dasein is left merely with its inner life. Heidegger says that the temporality that is profoundly boring "constitutes the ground of the possibility of the subjectivity of subjects, and indeed in such a way that the *essence of subjects* consists precisely in *having Dasein*, i.e., in always already enveloping beings as a whole in advance."[66] To be a subject is not to be an individual who is engaged in a determinate way in the world and who has an identity. The subject is an indeterminate "one" who is precisely not engaged with the world. Heidegger says that when we say that we are ourselves bored, we do not mean our individuated selves:

Yet we are now no longer speaking of *ourselves* being bored with . . . , but are saying: It is boring for one. It—for one—not for me as me, not for you as you, not for us as us, but *for one.* Name, standing, vocation, role, age and fate as mine and yours disappear. . . . This is what is decisive: that here we become an undifferentiated no one.[67]

In other words, one becomes a *subject.*

As I understand Heidegger, however, the contradiction at the core of profound boredom is that this indifference to everything is not complete. Dasein cares about this indifference and presumably it does not want to be bored to this extent. Dasein cannot settle for saying, "Nothing matters, so that does not matter either."[68] Simply shrugging one's shoulders and muttering "whatever" will only aggravate the problem. Like Kierkegaard's aesthete, profound boredom tries to exist as a contradiction by preoccupying itself with its inner life. This might be accomplished by what is currently called mindfulness, where one attempts to arrest the flow of time by focusing on the minutiae of each passing moment, trying to break it down into smaller and smaller parts in the attempt to hold onto it and to put off the inevitable moment when even that activity becomes boring.

Heidegger sums up the analysis of profound boredom to show that boredom is as ontologically basic as anxiety. Both can lead to an understanding of the whole of one's life as a coherent unity in the moment of vision:

Boredom is the entrancement of the temporal horizon, an entrancement which lets the moment of vision belonging to temporality vanish. In thus letting it vanish, boredom impels entranced Dasein into the moment of vision as the properly authentic possibility of its existence, an existence only possible in the midst of beings as a whole, and within the horizon of entrancement, their telling refusal of themselves as a whole.[69]

In other words, boredom eliminates the entrancement with the everyday world that leads to the forgetting of Dasein as an originary singularity.[70] But this vanishing leads Dasein to face up to its attunement and to take over explicitly the moment of vision for its own sake. Because the question of the meaning of our own lives can arise only with the recognition of the possible impossibility or the potential disappearance of everything, boredom is another pathway to authenticity in addition to anxiety. So although temporality is what is boring, it generates the "legitimate illusion" that "things are boring, and that it is people themselves who are bored."[71] Neither subject nor object, profound boredom makes possible the subjectivity of subjects. It shows that the essence of subjects consists in "*having Dasein*, i.e., always already enveloping beings as a whole in advance."[72]

Boredom and anxiety (and joy and awe) are thus moods that are revelatory of the whole of our being-in-the-world. Each of them also contributes to the temporality of Dasein in particular ways. At the end of the first division of *Being and Time* Heidegger has given a complete account of what it is to be a human being at a particular moment of time. In the second division, however, he wants to describe Dasein as a being whose life is "stretched out" in time between birth and death. His goal is to account for the "connectedness" of Dasein's life. This is both an ontological and a normative task. As an ontological task he needs to describe how Dasein can be the same being at different times of life. As a normative task he wants to show how it is possible for Dasein to fail to connect its life, on the one hand, and to succeed in integrating the various moments in a cohesive manner, on the other. To fail to connect is to be inauthentic, that is, not one's own, and to succeed in integrating one's life cohesively is to be authentic, that is, one's own.

Insofar as Dasein is always Mitsein or being-with-others, however, for the Dasein to be connected to itself, it must also be connected to its community. That is why it would be unsatisfactory if there were as many different times as there were individuals. Heidegger therefore owes us an explanation of the unity of time. As a provisional account, I would point out that as a member of a community and a generation, Dasein is initially constituted in a way that is "undifferentiated," that is, neither authentic nor inauthentic. Insofar as the Dasein is caught up in everydayness, for the most part it is inauthentic. Insofar as reckoning with time is necessary, the measuring of time can become a feature of our communal world. Clock time is thus a feature of the public, everyday world, and Heidegger claims that it is derivative from primordial temporality. Heidegger's account of Angst explains how Dasein can become authentic through resolve based on recognition of one's unavoidable finitude. It is important to realize that authenticity is not simply a function of the Dasein's connectedness to its past. Authenticity also involves Dasein's understanding of its present and its future. In fact, the past cannot be understood without understanding how it projects its future.

The future is the topic of chapter 4, where there is further discussion of the temporality of the distinction between the authentic and the inauthentic. At this point, we have Heidegger's account of the source of temporality and of normativity at hand. This account of the distinction between the authentic and the inauthentic allows for a provisional answer to the question of what Heidegger means when he says Dasein is time. Heidegger can be read as saying that Dasein *interprets itself* as temporal. Does this mean that Dasein could interpret itself as atemporal? The answer is no, not if to interpret itself means that Dasein exists *as* its interpretation. But if Dasein can only interpret itself as temporal, is Heidegger's claim vacuous? Again, the answer is no, because there are at least two possible ways in which Dasein can exist temporally, namely, authentically and inauthentically. Heidegger's claim is thus not vacuous. On the contrary, it makes all the difference to our lives. How the normative is reflected in each of the temporal modes of past, present, and future can now be discussed in detail in the next chapters.

Reflections

To sum up the results of this chapter, let me review the various answers our philosophers have given to the question, what is the source of time? From an initial reading of Kant's first *Critique*, especially the Transcendental Aesthetic, it would be fair to conclude that his answer is that the source of time is the mind. As the form of intuition, it would seem that time is sufficiently mind dependent for us to be able to say that without mind there would be no time.

Heidegger's reading of Kant in *Kant and the Problem of Metaphysics* in 1929 specifies the source of time in Kant more precisely as the faculty of imagination. Through his analysis of the section of the A edition called the Schematism, Heidegger was able to see the transcendental imagination as the spontaneous welling up of the temporal. Heidegger then went on to attribute to Kant his own speculation, which was that the mind did not produce time so much as time produced the mind.

In the meantime, Heidegger's mentor Husserl had lectured on internal time-consciousness between 1905 and 1910, and Heidegger had edited and published a version of these lectures in 1928. Although Kant offered an analysis of Newtonian, objective time, he did not have a specific theory of lived temporality. Husserl was the first to provide an account not so much of time as of time-consciousness. His introspective method of phenomenology led him to posit such time-consciousness as "inner." Husserl's contribution was intended to go beyond Kantian faculty psychology whereby time was imposed on the data of sensation by a faculty in the form of a synthesis that produced experience. Instead, he located duration in the experience of the moment by saying that each moment was not isolated, but connected both to the previous moments through retentions and to future moments through protentions. Once again, however, the source of temporality was taken to be internal and subjective.

Heidegger's own account of temporality requires the source of temporality to be neither subjective nor objective. Instead, temporality is itself the source of the subjective–objective distinction. What does this mean? To put the point in a formulaic way, we could say that in contrast to Kant's view of time as mind dependent, Heidegger's view is that it is the other way around, and mind is time dependent. In Heidegger's terms, the statement that what temporalizes is temporality itself is intended to avoid reifying the mind into a present-at-hand object. Temporality has to do with the way that comportments occur. An important aspect of such behavior is attunement, which is reflected in moods, feelings, and emotions more than in explicit self-conscious cognitions. As we saw, the method that Heidegger uses to characterize moods such as boredom is called "awakening." Unlike Husserl's method of phenomenological reduction, which brackets the reality of the subject and object to focus on consciousness per se, Heidegger believes that it is important to get down below the level of consciousness to the phenomena themselves. Making some aspects of our lives explicit tends to distort or destroy them. Heidegger calls this derivative way of bringing things to explicit consciousness "ascertaining." I am suggesting that Heidegger sees Husserl's phenomenological method as a form of ascertaining, in contrast to Heidegger's own method of awakening.

Awakening reveals how the source of time is in temporality, and the source of temporality is nothing other than temporality itself. Heidegger's elaborate example of such awakening is his discussion of boredom. Insofar as he was able to reduce the 150 pages of lectures to one published sentence in the essay "What Is Metaphysics?," it would seem that the basic idea is not all that difficult. The point is that objective time is dependent on lived temporality and that the reverse is not the case. From that point of view, the way to describe temporality is not to reduce it to something else, but to see how it shows up in our implicit encounters with the world.

There are several other philosophically interesting issues or ideas that have come up in the course of this chapter that I would like to highlight. One of these is the contrast between explanations of temporality through faculty psychology and through accounts of duration. If Kant is the paradigm of the former, Husserl and, as we will see, Bergson, are the quintessential theorists of the latter. Where to fit Heidegger into this distinction is not so clear. On the one hand, his notions of human existence as divided into three major aspects of comportment—*Befindlichkeit* (disposedness), *Verstehen* (understanding), and *Rede* (discourse)—bear a certain resemblance to faculty psychology. His analysis of temporality into the three *ecstases* of past, present, and future (discussed in later chapters), however, is more in the tradition of duration theory. He was certainly aware of both Husserl and Bergson, as well as earlier philosophers of time from Aristotle and Augustine to Kant and Brentano. In general, then, as we encounter other philosophers of temporality in the following chapters, it may be productive to ask under which paradigm each of them figures. Then we can ask how they would solve the problem of explaining how time is one if temporality is relative to each individual. Is temporality so local that there are as many different times as there are people?

Another set of issues arose in this chapter around the idea of subjectivity, and they will need further investigation.[73] For Heidegger the idea of subjectivity may be what Robert Brandom has dubbed a "Bad Idea," one that should be dropped because of all the philosophical baggage that goes along with it. Or perhaps a more moderate approach would be to say that of course people have access to their own experiences, but that one should not try to build a philosophical method of phenomenological reduction around this minimalist claim. For Heidegger, under this construal, subjectivity would not be interesting to the philosopher, since it is a derivative and ontic mode, one that has some everyday use but no special philosophical significance. Insofar as it designates a derivative mode of experience, its emergence can be explained by more basic phenomena such as boredom or anxiety. In other words, subjectivity is to be explained; it does not do the explaining. More interesting will be ideas like the self and the individual, which are not identical to the idea of subjectivity. In relation to Heidegger, for instance, one might well ask, if Dasein is not first and foremost a subject, what is it? What does "Dasein" refer to exactly?

Furthermore, there is a set of problems about the relation of subjectivity and self-consciousness. Michel Foucault, for instance, gives us a method for describing how subjects are formed by social practices before they are self-conscious of who they have become. Moreover, there is not simply one form of subjectivity exemplified by all subjects, but different subjectivities are formed under different "cultural politics."[74]

This reference to social practices and cultural politics raises issues about the possibility of phenomenology not simply describing experience, but also prescribing normativity. At this point we have one account of the birth of normativity, namely, Heidegger's use of temporality to explain the distinction between authentic and inauthentic comportments. Later discussions will focus on the pertinence of this account of normativity to political and ethical attitudes toward the past and the future.

2 There Is No Time Like the Present! On the Now

Although clichés about time generally sound like truisms, they can also be revealing. In this particular case, "there is no time like the present" is often used as a practical adage. "Now is a good time to take action" would be another way of stating this advice. In this sense, it is closely linked to "Carpe diem"—seize the day! The expression can also be viewed not as practical advice, however, but as an ontological claim, one that points to the reality of the present and the unreality of the past and the future. "There is no time *but* the present" would be a better way of making this ontological point. So stated, the expression captures the common intuition that the present obtains in a different way than the past or the future. Some philosophers, however, have the contrasting intuition that the present takes no time at all and that ultimately there is no such thing as the present. Augustine raised this issue by asking whether the present is so instantaneous as to be practically nonexistent. Then if the past is gone, and the future is always not-yet, what does that say about the reality of time in general and of the present in particular?

In this chapter on the present I do not intend to tackle directly the metaphysical issues about the reality of time. Instead, I continue to approach the issues through the phenomenology of temporality. By starting from the analysis of temporality—from the time of our lives—phenomenological philosophers expect to avoid metaphysical issues about the reality of time. Phenomenologists tend to think that objective time is real, but they see it as derived from the more primordial way in which humans temporalize their world. Hermeneutical phenomenologists then add that temporalization is a basic form of interpretation. Interpretation in the broad sense is not the result of self-conscious, reflective theorizing, but is built into the activities and projects in which humans are engaged.

These strategies for talking about the time of our lives as opposed to the time of the universe will emerge from this chapter's analysis of the temporality of the present. The chapter begins with Hegel, focusing in particular on his brief but historically influential critique of the Now as a form of sense-certainty. I then discuss the concerns about the size of the present voiced by William James. Both Heidegger and Derrida portray Husserl and Merleau-Ponty as paradigmatic phenomenologists who therefore become the targets for the deconstruction of the phenomenological notion of presence. Nietzsche appears at the end because his account of eternal return represents an entirely different theory of the present. Each of these thinkers will appear again in later chapters that deal with the other temporal dimensions and the particular philosophical problems associated with them.

Hegel's Critique of the Now

What is the present? One metaphor handed down to us from antiquity construes the present as the boundary between the past and the future. Insofar as this boundary has no duration, the present is dissolved by the skeptical intuition that the instant is over before one knows it and is indeed nothing at all. In the modern tradition, Hegel raises similar skeptical issues about the reality of the present in the first chapter of *Phenomenology of Spirit*. Here he attacks the position of sense-certainty for relying on the idea of the Now as an unquestionable item of knowledge. Although we think that we know with certainty that it is Now, and thus what the word "Now" refers to, in fact, it is difficult to articulate what it is that we think we know. More technically, although "Now" seems like an indexical term referring to a bare particular, it can also function as a universal that refers to any and every moment. Hegel argues that if I write down "Now it is morning" and then look at the sentence at night, I will see the certainty of the term vanish.

A typical reaction to this argument is that it would have been more convincing if it did not rely on writing down the sentence. Could one not simply change its tensed status by attaching a date and place stamp to the writing? For instance, by writing "Now, at 7:15 AM in California on such and such a date, it is morning," the sentence would then always be true. This response misses Hegel's point, however. Hegel is making a phenomenological claim that emphasizes the importance of the observer in determining the temporal sequence of events. Hegel's dialectical strategy is to show that even if the temporal sequence seems to exist in itself, without a fixed standpoint to contrast to the flux of experience there could not be any before or after, earlier or later, faster or slower. The time that seems objective and independent turns out to be dependent on subjectivity.

Hegel has two other arguments that supplement his attempt to problematize the Now. The first concerns the fleeting character of the Now. Whenever I identify myself as having an experience right now, that moment is already over, and the Now is already in the past. If this were right, then one could never use the term "Now" to refer to the present moment. The Now to which one intended to refer would never be the Now that was actually occurring. In Hegel's words, "The Now that *is*, is another Now than the one pointed to."[1]

The second issue concerns the divisibility of the Now. Hegel thinks that any Now "contains within it many Nows."[2] When I say "Now," therefore, that to which I am referring is not obvious. I could be referring to today. A day contains many hours within it, however, and an hour includes many minutes, and minutes include seconds. I could be referring to anything from a few seconds to several decades. These two problems are said to show that what the term "Now" refers to is not as clear as sense-certainty thinks. Hegel thus problematizes the naive intuition that time is objective and mind independent by deconstructing the notion of time as consisting of instants.

William James and the Specious Present

In *The Principles of Psychology*, William James continues the tradition of skepticism about the reality of the present in order to establish not only that temporality is dependent on the observer, but also that time-consciousness and memory are not the same. He is more interested in consciousness than in time per se, and is thus focusing more on what I call "temporality" or "lived time." His analysis of the perception of temporality begins by considering two strongly skeptical views on the nature of time-consciousness. First, James derides what could be called the "glow-worm" theory of consciousness. On this theory, each moment of consciousness is separate from every other moment of consciousness. James cites a contemporary text in which the view is described as follows:

One idea, upon this supposition, would follow another. But that would be all. Each of our successive states of consciousness, the moment it ceased, would be gone forever. Each of these momentary states would be our whole being.[3]

James suggests that consciousness on this view would be "like a glow-worm spark, illuminating the point it immediately covered, but leaving all beyond in total darkness."[4] He maintains that it is doubtful that a practical life would be possible under these conditions.[5]

On the glow-worm theory, the present is the only moment of time that really obtains. On another view, not only do the past and future not exist, but neither does the present. James cites the following speculations by S. H. Hodgson:

Crudely and popularly we divide the course of time into past, present, and future; but, strictly speaking, there is no present; it is composed of past and future divided by an indivisible point or instant. That instant, or time-point, is the strict *present*. What we call, loosely, the present is an empirical portion of the course of time, containing at least a minimum of consciousness, in which the instant of change is the present time-point.[6]

James then urges his readers to ask themselves whether they can really introspect the present. His conclusion draws on the limitations of reflective introspection of phenomenal experience: "Reflection leads us to the conclusion that it [the present] *must* exist, but that it *does* exist can never be a fact of our immediate experience."[7] James is alleging that ordinary common sense therefore commits the fallacy of deriving an objective *is* from a subjective *must*. Simply because someone *thinks* that something *must* be the case does not entail that it *is* the case. This mistake is quite common, especially in drawing conclusions about consciousness from introspection.

James's own attitude toward the present is different from both of these views, but as I understand him, it is still a skeptical view. He borrows the term "specious present" from one of his contemporaries, whom he cites as follows: "The present to which the datum refers is really a part of the past—a recent past—delusively given as being a time that intervenes between the past and the future."[8] He believes that the term "Now" equivocates between a knife's-edge and a saddleback conception of the present. The former thinks of the present as an instant, roughly equivalent to the snap of one's fingers. The latter assumes that the present itself takes time and that it lasts for a while. The present is thus ambiguous insofar as it connotes both instantaneity and duration.

James then makes some intriguing observations about the phenomenology of time-perception in its relation to memory. He suggests that temporal duration is not only stretched out, but is also directional. The duration has both a bow and a stern, he says. But we do not experience first one end, then the other, until finally, "from the perception of the succession [we] infer an interval of time between."[9] Instead, he maintains that prereflectively we feel "the time interval as a whole, with the two ends embedded in it."[10] Reflection may result in "decomposing" the experience into its beginning and its end. Time-perception, perhaps unlike time itself, is a synthesis of the two directions. Whereas metaphysically, time may be simple (i.e., it cannot be divided any farther), James suggests that time-perception is a synthetic datum.

The specious present becomes, for James, the primordial unit of time-perception. He emphasizes that "the original paragon and prototype of all conceived times is the specious present, the short duration of which we are immediately and incessantly sensible."[11] The specious present allows a being to distinguish before from after, and thus to have a sense of time's directionality. However much the content of consciousness varies (and it is constantly in flux), the specious present is a permanent framework, "with its own quality unchanged by the events that stream through it."[12]

One might think that the fading away into the past means that memory must be involved. James asks himself whether a being that did not have memory could still have a rudimentary perception of time. For James the answer is yes, at least if he is right to think that the experience of the specious present as fading into the past is different from memory. Memory brings back or "reproduces" an event that has completely faded out. The immediate past that is part of the specious present is thus different from a remembered past. James distinguishes between the retained past of the specious present and the remembered past of memory. In James's colorful language, he asks his readers to observe "that the reproduction of an event, *after* it has once completely dropped out of the rearward end of the specious present, is an entirely different psychic fact from its direct perception in the specious present as a thing immediately past."[13] The immediate past that is part of the specious present is thus not the past as remembered. The remembered past is the entire unit of what was once the present. The immediate past is only a part of the experience of the specious present. When memory recollects a present that is now past, that past present will include its own sense of what was for it the immediate past. That is why James thinks that a being with no memory could still have a sense of time. Such speculations lead me to interpret James as saying that temporality is a necessary but not a sufficient condition of memory, and memory is a sufficient but not a necessary condition for temporality.

If this formulation is correct, the next question is exactly how long does the present last for lived time? James knew, of course, about attempts by psychologists to determine the range of the perception of the Now. This range would be between the smallest amount of time that can be perceived and the longest amount of time of which we could be said to be immediately aware. In James's time, the shortest time would be expressed in thousandths of a second. Today we can measure in even shorter spans of nanoseconds and attoseconds, which is perhaps why we think that attention spans are getting shorter. At the other end of the scale, James tends to put the upper limits of the present at approximately a dozen seconds, although in his conclusion he specifies the duration of the specious present as "varying in length from a few seconds to probably not more than a minute."[14]

James's guess is not too far from present-day research. Neuroscientist Ernst Pöppel speculates that "we take life three seconds at a time," three seconds being the time it takes for a handshake, short-term memory formation, preparation for a golf swing, formation of speech phrases, or pauses while channel surfing.[15] However long it lasts, the specious present always includes both the warm-up and the fade out. These two aspects make temporality what it is, and both are built into time-perception right at the start. James thinks of the neurons not as digital switches that are either on or off, but as analog relays with a fading effect, much like a radio tube or a lightbulb that glows for a while after being shut off.

James explains the feeling of duration by arguing, on the one hand, that the "feeling of past time is a present feeling,"[16] and on the other hand, that the moment that is experienced as present is already fading into the past. (In a footnote he suggests that in addition to *fading* brain-processes, *dawning* processes contribute equally to the feeling of duration in the specious present.[17]) In other words, the sense of the past is built into each specious present because the specious present itself is experienced as already fading into the past. James infers from this phenomenon that if we tried to imagine Adam's first experience, we would realize that there could not be such an experience. James emphasizes his position by asserting that "the new-created man would unquestionably have the feeling, at the very primal instant of his life, of having been in existence already some little space of time."[18] The Adamic first experience is a myth, as is Adam himself.

Husserl on Time-Consciousness

Husserl's *Zur Phänomenologie des inneren Zeitbewusstseins*[19] includes lectures that Heidegger edited and published in 1928— toward the end of a decade during which Heidegger was himself writing intensively about time. The most recent translation, by John Barnett Brough, contains many notes and drafts that Husserl never published and that do not form a single work. The task for Husserl scholars is, then, to determine how his thinking evolved, and to use these hypotheses to date the various jottings in order to determine which of them represent his more considered views.[20] This account of duration is highly complex, as a quick look at figure 2.1, one of Husserl's many graphs of temporality, will indicate. Nevertheless, the basic idea represents a historical advance over William James's account, and I will present it using a minimum of technical vocabulary in order to bring out the intuitive appeal of Husserl's model.

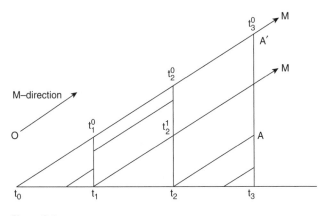

Figure 2.1
Husserl's time graph. Redrawn from Edmund Husserl, *On the Phenomenology of the Consciousness of Internal Time (1893–1917)*, trans. John Barnett Brough (Dordrecht: Kluwer, 1991), p. 343.

Husserl is concerned, as James is not, to distinguish between the empirical analysis by psychology and the supposedly a priori features discovered by phenomenology, construed as transcendental philosophy. A *transcendental phenomenology* would reveal the a priori structures of consciousness. Both Husserl and Heidegger think of philosophy as raising ontological or transcendental questions that empirical psychology presupposes but cannot raise by itself. One such question is, which comes first, experienced temporality or objective time (i.e., clock time)? Another issue that arises is how can there be *one* time if everybody has different temporalities? A third question is to ask whether temporality can be said to "flow." If so, in what direction does it flow, and what happens to the present, which is always there even though the content is always different?

The phenomenological program of analyzing temporality differs dramatically from the Kantian style of explanation of time through faculty psychology. Faculty psychology constructs experience as the outcome of a transcendental processing machine of which we have no awareness. Somewhere in the process, time is added to experience by one faculty or another so that we can distinguish between past and present experience. This explanation of how we order our experiences of time does not account, however, for the temporality of the experience as such. There is no account of *duration*. Duration involves the qualitative aspects of temporal experience, whereas faculty psychology confines its concerns to explaining the orderability of experience and thus the quantitative aspect of time measurement required by Newtonian science.

Husserl's phenomenology is conceived as a description of what Husserl calls "intentionality." Intentionality is simply consciousness of something. A special kind of intentional object is a temporal object. Husserl's favorite example of a temporal object is a melody. Melodies have what we call duration. Husserl's phenomenological task is thus to describe duration, which is a particular kind of intentional experience, different from James's specious present.[21] If temporality is a flow, then duration is a feature of the flow.

Right at the start, Husserl can be interpreted as dismissing Kant's efforts to find the permanent, without which Kant thought that one could not even speak of change. Husserl also appears to discount Kant's efforts at refuting idealism and establishing the permanence of external substance when he writes:

Where is the object that changes in this flow? Surely in every process a priori something runs its course? But here nothing runs its course. The change is not a change. And therefore it also makes no sense to speak of something that endures, and it is nonsensical to want to find something here that remains unchanged for even an instant during the course of a duration.[22]

With the assertion that "the flow of consciousness constitutes its own unity,"[23] Husserl moves beyond faculty psychology with its transcendental machinery for unifying something logical (concepts) and something nonlogical (intuitions). In Kant's case, the opposition between concepts and intuitions is the principal obstacle that has to be overcome. Husserl's explanation of duration and the phenomenon of flow bypasses and obviates a significant part of Kant's transcendental philosophy.

As for temporality, if one's theory of perception holds that temporality is such that one can hear only what is in the present, and the present can include only one note at a time, then one would never hear the melody. The melody is the entire sequence of notes, including their length and the space in between the notes. For instance, how does one otherwise hear the rhythm? On Husserl's account, the notes that have already been played are retained, and *retention* includes the quality of "sinking" into the past. Even when one hears the same note over time, what Husserl calls *primary memory* is involved. That is because although one is hearing the same note and therefore is having identical sensory input at the different moments throughout which the note is maintained, there is still a difference insofar as the beginning of the tone is different from the middle or end of the tone. As Husserl says in §31, there is a new "primal impression" that corresponds to each new Now.[24]

Husserl's contribution to the history of the phenomenology of temporality is to make a sharper distinction than previous philosophers between time as the noematic or objective correlate and temporality as the noetic correlate, that is, consciousness or experience. Temporality involves a three-layered phenomenon of the primal impression, the protention, and the retention. The protention is the projected horizon, the intentional anticipation that reaches toward the immediate future, just as the retention holds onto the immediate past in the fading out of the primal impression.[25] Past, present, and future are different from retention and protention. We experience ourselves as in time and as having a past, present, and future *because* our temporality involves the structure of protention, retention, and primal impression. Moreover, protention, retention, and primal impression are all part of each experience. Unlike past, present, or future, a person could not have an experience of only one of these aspects of temporal experience. It takes all three for one unified experience to be possible.

Although Husserl has much more to say about retention than protention,[26] the central claim is that temporal experience is not like a string of pearls. In a string of pearls each pearl is self-contained and identifiably discrete from every other pearl in the string. At the same time, in a matched string each pearl resembles every other pearl such that they all look alike. Husserl's picture of duration is a spatial diagram in which each moment reflects the previous one and anticipates the next one. The string of moments of time constitutes itself from within its own structure rather than being the result of an external synthesis by a faculty of the mind.

Husserl's struggle to explain retention was a long, drawn-out affair. By 1911, however, he felt he had a solution: "There is one, unique flow of consciousness (perhaps within an ultimate consciousness) in which both the unity of the tone in immanent time and the unity of the flow of consciousness itself become constituted at once."[27] To elaborate on his example, imagine listening to an opera in which the soprano hits and holds an impossibly high note. As you wonder how long she will be able to hold it, you are aware of how long she has been holding it. How would that awareness be possible? You know it is the same note, yet you also know that it is stretched out over time. Husserl thinks that you retain the initial sound throughout its duration so that you then know that she held it for an extraordinarily long time. In fact, at each instant that she is holding the note, you are retaining not only the prior instant but also your retention of the prior instant before that. Figure 2.1 shows how complex this interlacement of retentions can become.

The complexity of the retentions of retentions of retentions strikes Husserl's critics, including myself, as in fact a problem with his account. One Husserlian tries to defend Husserl by saying that "it is by remembering the past perception that one is able to recall the past object of the perception."[28] This claim seems to get the phenomenology wrong. I do not first remember my memory and then remember the object. Furthermore, I doubt that it is right to say that I can retain a retention of a retention. Just as I do not perceive my *perception* of an object, but the object itself, I do not remember having a memory (whether primary or secondary), but I remember the content of the memory first. Only then can I remember having remembered it on earlier occasions.

If I am right about the phenomenology, then it is not surprising to find Husserl himself saying much the same thing about both perception and memory. In §27 Husserl says that in order to have perception, which "constitutes the present," I do not represent the perception; "rather I represent the *perceived*, that which appears as present in the perception."[29] I read this as saying that we do not *perceive* the perception but the object. He then specifies,

Memory therefore does actually imply a reproduction of the earlier perception, but the memory is not in the proper sense a representation of it: the perception is not meant and posited in the memory; what is meant and posited is the perception's object and the object's now, which, in addition, is posited in relation to the actually present now.[30]

Thus, I understand Husserl as saying as well that I do not *remember* the perceiving so much as the perceived object.

Although Husserl's account of retention is more detailed than William James's view, it does not apply quite as well to protention. One might grant that the present lingers in our perception as it passes into the past, but can we say as a corollary that we see into the future as the protention becomes present? Husserl may have thought the situation for the past and the future were parallel. Insofar as we act, we act *toward* some end, and this slight leaning ahead of oneself could be seen as a correlate of the slight lingering of one tone as another succeeds it. If we apply James's metaphor of the ship, Husserl's analysis would have the prow of the present cutting into the future as its stern slips into the past.

An obvious objection, however, is that this analysis breaks the ship in two. To deal with this problem, Husserl distinguishes two perspectives that can be taken on the present and its flow. What he calls the "horizontal intentionality," or better, "longitudinal intentionality" (*Längsintentionalität*) is distinguished from "transverse intentionality" (*Querintentionalität*).[31] This distinction is both tied to Husserl's account of how self-awareness is possible and also involved in the explanation of how temporal passage is able to be experienced. The *Längsintentionalität* concerns the stretch of the temporal series. The *Querintentionalität* then would be a transverse view across the stretch. Although he does not say so, the distinction is described in a way that resembles the structuralist distinction between the synchronic and the diachronic, that is, the oneness of temporality at any given moment and the stretch of time over its many moments. This interpretation is supported by the fact that he uses the term *Momentan-Zugleich* or "momentary being-all-at-once" in just this sense of the synchronic, although he notes in the margins that he really means the diachronic *Strecken-Zugleich* or "the stretched being-all-at-once at length."[32]

To illustrate his account, we can return to his example of hearing a melody. When one listens to the note for its own sake (perhaps because it is slightly flat or perhaps because it is remarkably pure), one is making the temporal slice across the flow of time that we call the present. But hearing each note, one after the other, is not the same as hearing the melody. The melody requires a stretch of connected time. Thus, when one is hearing the last resolving chord in, for example, the famous opening phrase of Beethoven's Fifth Symphony—da-da-da-*dum*—the first note is over when one hears the last note. But here Husserl's account of retention, as described above, is intended to save the day with this analysis of retention and protention, which explains how the series of notes is heard as a unity, that is, as a melody. The "all-at-once" or *Zugleich* brings out how the oneness at a given moment and the unity of a "stretched" temporal object like this are constituted.

For Husserl, internal consciousness involves a prereflective awareness of the temporal ordering of the experience itself. What this awareness is an awareness of is the *self-givenness* of the primal impression, that is, the sense one has that one is being presented with some content. The content of each primal impression will vary, but the self-givenness of the experience can be distinguished from this content. The Husserl scholar Dan Zahavi maintains that to *distinguish* the content and its self-givenness does not mean that the self-givenness can be *separated* from the content.[33] To separate the givenness and the content would make the givenness into a separate datum, but that gets the phenomenology wrong. There is no empty field into which the contents flow, and the stream of consciousness is not itself something that can be made into a reflective content of consciousness.

Maurice Merleau-Ponty has worked out the most detailed analysis of the prereflective level of experience. Insofar as Merleau-Ponty sees himself as working out Husserl's account of *Zeitbewusstsein*, we can continue the discussion of Husserl by taking up the views of his most influential French interpreter. An equally important antecedent for Merleau-Ponty's thinking about the present is, however, Heidegger. Before moving on to the French phenomenologist, therefore, let us look more closely at Heidegger's views in *Being and Time* about the present.

Heidegger in *Being and Time*

Martin Heidegger's contribution to the history of the phenomenology of temporality is double edged. He has to criticize Kantian faculty psychology and at the same time explain more convincingly than Husserl how objective time relates to temporality and the experience of duration. Although faculty psychology might succeed in explaining the constitution of time as a series of discrete units or Nows, at least three respects have been mentioned in which it does not constitute a proper *explanation* of temporality. First, faculty psychology is empty. It tries to explain how something occurs by telling us only *where* it occurs. Second, faculty psychology is circular. Just as sleeping pills would not be "explained" by saying that they have dormative powers, temporality would not be explained by ascribing it to a faculty that is said to have the function of injecting time into the synthesis of experience. Third and most important, it leaves out a fundamental phenomenological feature of temporal experience, namely, duration. Husserl's critique of his more Kantian predecessor Franz Brentano is important because it leads to a better explanation of duration (although whether it is the *best* account of duration will have to wait until chapter 3 and a comparison of Husserl and Henri Bergson). Heidegger has to find a more basic level for analysis than either Brentano or Husserl envisioned.

Near the end of *Being and Time* Heidegger makes a three-way distinction between primordial (or originary) time, world time, and ordinary time. His goal is to show that time is better explained by starting from primordial time rather than from the ordinary understanding of time. There are three features of time that the ordinary understanding of time does not capture or expresses incorrectly. First, for reasons explained below, time has an irreversible directionality to it that the ordinary understanding of time cannot explain. Second, time is finite, but the ordinary understanding views it as infinite. Third, the metaphor for time as a river is wrong because time does not flow "downstream," as it were. In particular, it does not flow from past to the present and toward the future. Heidegger's thesis is that starting from the ordinary understanding of time leaves these phenomenologically determinable features of temporality unexplainable. These phenomenological features can be understood only by starting from primordial time and showing how ordinary time is derived from it.

Time is explained differently if it is understood as ordinary time, world time, or primordial time. The ordinary understanding is that time is an infinite series of Nows, that is, countable, discrete points that succeed one another in a sequence, much like a string of pearls. This ordinary understanding of time makes the mistake of turning time into moments, that is, a series of present-at-hand entities.

Deeper than ordinary time is world time, which has the phenomenological characteristics that Heidegger calls datability, significance, spannedness, and publicness. Datability *assigns a time*, such as "now," "then," or "on that former occasion."[34] Spannedness adds what Bergson would have called "duration," except for Heidegger's charge that Bergson still sees the problem in terms of exteriorizing subjective, qualitative temporality. For Heidegger, the span of time can vary with the interpretive range of what is significant. Humans find themselves thrown into a situation that already determines what is significant and what is not. Because Dasein is always a being-with-others, time will have a public character insofar as it is used to coordinate different activities.

Temporality is thoroughly interpretive, and there is no single way of correctly assigning time. "The interpretative expressing of the 'now,' the 'then,' and the 'on the former occasion,' " writes Heidegger, "is evidence that these, *stemming from temporality, are themselves time.*"[35] Heidegger shows the importance of interpretation in understanding temporality through a play on the verb for "to interpret." He writes, "The making-present which awaits and retains, lays 'out' [*legt . . . 'aus'*—a play on '*auslegen*,' to interpret] a 'during' *with a span*, only because it has thereby disclosed *itself* as the way in which its historical temporality has been ecstatically *stretched along*, even though it does not know itself as this."[36] Temporality is interpretation with a lower-case "i" (in German, *Auslegung*) as opposed to Interpretation with an upper-case "I" (in German, *Interpretation*). The distinction is between the prereflective ways in which Dasein copes with its world and the more reflective or conceptually articulated Interpretation that formulates its commitments in words.[37]

There can be authentic and inauthentic interpretations of temporality. One inauthentic interpretation, which Heidegger refers to as the "lost Present," is illustrated by Heidegger's analysis of people

who never have enough time. They are often late because of the
need to dash from one distraction to another.[38] These people have
lost or wasted time because they have lost themselves in the busy-
ness and distractions of everydayness. Heidegger does not explain
the word "lost," except to use it in his brief account of falling into
the Present. He describes the "lost Present" as the "leaping away"
of the Present from both its authentic future and its authentic having
been.[39] This movement results in "*taking a detour*" through the
Present. Or, instead of speaking of taking a detour, in current parlance we could draw on the idea of "getting lost" and say that one
is "losing it." As Heidegger says, "The 'leaping away' of the
Present—that is, the falling into 'lostness'—has its source in that
primordial authentic temporality itself which makes possible
thrown Being-towards-death."[40]

In contrast to this lostness in the everyday, authentic Dasein
always has time and is always on time. For the authentic Dasein,
time passes, but it does so as a coherent connectedness rather than
as a disconnected leaping from one missed opportunity to the next.
Instead of letting the past take over the present, the authentic relation to the present involves the *Augenblick*—the "moment of
vision" that is gained in the "glance of the eye." In this moment of
vision Dasein does not lose sight of the present but instead gains
it by projecting for itself a unified vision of the connectedness of
its past, present, and future. Dasein then resolves from there on out
to act consistently on this momentous insight into the coherence of
its own life.

Heidegger's distinction between the authentic and the inauthentic shows that Dasein participates in the temporalization of its
own life. Insofar as both an authentic and an inauthentic relation
to the present are possible, temporality (unlike objective time)
is not necessarily successive or consecutive. An inauthentic life
is an example of a disconnected temporalization. Heidegger's
emphasis on Dasein's own role in creating itself as a unified self
stretching from birth to death shows that he values coherence over

discontinuity. Although poststructuralists such as Foucault or Derrida may challenge this normative assumption of the value of narratival unity, such a challenge should not obscure the more basic point, which is that a life can be temporalized in different ways by the Dasein. Of course, as Heidegger's discussion of history in *Being and Time* brings out, we do not have control over our own personal fates or the destiny of our community. Nevertheless, individually we are responsible for our own particular temporalizations of our existences.

To return to the ordinary understanding of time, there is a contradiction between two common ways of thinking about time, namely, as a staccato of disconnected Nows on the one hand, and on the other hand, the sense of time passing as a coherent flow. Time is often construed as a river in at least two senses: first, because time "flows," and second, because it goes in only one direction. For Heidegger, however, temporality is neither staccato nor fluvial. Temporality has more connectedness and stretch than a staccato of Nows would have. Moreover, temporality is not fluvial because it is not experienced as flowing from the past into the present and toward the future. On the contrary, for Heidegger the future has priority. Temporality comes out of the future, and then it goes into the past and comes around into the present. For Heidegger the inauthentic present attitude is to sit back and wait for time to pass and for things to happen. In this way, one just blunders along without any focused attempt to connect one's life. In contrast, in the authentic present attitude of the *Augenblick*, I project a meaningful course of action and I resolve, without relying on any external or extraneous input, to live my life in a coherent and connected way.

The ordinary way of thinking about time gives a priority to the present, but to one that is lost. In contrast, primordial temporality emphasizes the future. Both "present" and "future" mean something different, however, on the different understandings of time. For the ordinary understanding, time always appears as the Now. The past would then consist of formerly present-at-hand Nows that *have* occurred, and the future would consist of soon to be present-at-hand Nows that *will* occur:

> Thus for the ordinary understanding of time, time shows itself as a sequence of "Nows" which are constantly "present-at-hand," simultaneously passing away and coming along. Time is understood as a succession, as a "flowing stream" of "Nows," as the "course of time."[41]

The ordinary conception of time makes the mistake of separating the temporal into separate domains, depending on the status of the Nows, whether they are over, yet to come, or actually occurrent. Of the ordinary present Heidegger writes, "In the way time is ordinarily understood, however, the basic phenomenon of time is seen in the '*Now*,' and indeed in that pure 'Now' which has been shorn [*beschnitten*] in its full structure—that which they call the 'Present.' "[42] He describes the ordinary conception of the future as "a pure 'Now' which has not yet come along but is only coming along."[43] Similarly, the ordinary sense of the past is as "the pure 'Now' which has passed away."[44]

Given this distinction between the ordinary understanding of time and the primordial understanding of temporality, what giving priority to the present or the future or the past means will differ. The ordinary conception of time simply differentiates those Nows that have occurred from those that have not yet occurred. Heidegger is right that this is an unsatisfactory account of time. Heidegger says, "When Dasein is 'living along' in an everyday concernful manner, it just never understands itself as running along in a Continuously enduring sequence of pure 'Nows.' "[45]

The argument for this criticism of the ordinary conception of time depends on seeing that if the Now were the most basic unit of time, that conception would not suffice to distinguish those units that have been from those that have not yet occurred. For one thing, this account presupposes rather than explains time. To say that the past consists of Nows that are past is unhelpful at best (and circular at worst).

Furthermore, one cannot take a Now (a temporal moment) and read from its face whether it is past, present, or future. These dimensions are not perceivable in the examination of the moment, but their status is purely relational. That is to say, whether a moment is past or to come depends on its relation to other moments, and there is nothing intrinsic to the moment that tells whether it is past or futural. Heidegger thus thinks that the directionality of time cannot be explained by the ordinary understanding of time: "Why cannot time be reversed? Especially if one looks exclusively at the stream of 'Nows,' it is incomprehensible in itself why this sequence should not present itself in the reverse direction."[46] In other words, there is something missing in the account of time that starts from the Now. Heidegger suggests that what is missing are datability and significance:

In the ordinary interpretations of time as a sequence of "Nows," both datability and significance are *missing*. These two structures are *not* permitted to "come to the fore" when time is characterized as a pure succession. The ordinary interpretation of time *covers them up*. When these are covered up, the ecstatico-horizonal constitution of temporality, in which the datability and the significance of the "Now" are grounded, gets *leveled off*. The "Nows" get shorn [*beschnitten*]of these relations, as it were; and, as thus shorn, they simply range themselves along after one another so as to make up the succession.[47]

That time is datable means that it is connected to my practical activities such that different times of day have different significances for me. Time is not simply the bare numbers, but is tied to the coordination of events (for instance, the need to be in class by 1:30 PM or at dinner by 8:00 PM).

The main point of this analysis is that no matter how the directionality is described, whether flowing from the past into the future or from the future into the past, there will be an irreversible direction for temporality. If the direction of objective *time* is reversible and therefore indifferent for quantum physics, *temporality* is not reversible for us. There will necessarily be an experiential difference between that which has already occurred and that which has not yet occurred. From the phenomenological standpoint, any theory positing the reversibility of time or the possibility of time travel will be false by reductio. If time seems unimportant for the physical sciences, where it appears only occasionally in the formulations of laws, temporality should be at the center of philosophical concerns, for everyone must come to terms with its passing.

In contrast to the failure of the ordinary understanding of time to capture the phenomenal difference between the past, present, and future, Heidegger's phenomenological account is more successful. Drawing on Husserl's discussion of how each Now involves both retention of previous Nows and a protention of futural ones, Heidegger views the Now not as a self-contained moment the way that the ordinary conception understands it metaphysically. Instead of discrete Nows, for Heidegger authentic temporality involves what he calls "ecstases" whereby the present is really an anticipation of a future from the standpoint of an already configured past. The future has priority over the present because the present experience is of a future that is coming toward the present. There is no pure presence because the present that is experienced is always already past. The experience of time passing is not the flow of time from a past toward a future so much as of the future coming into the present. In a reorientation of James's account of Adam, whereby there is no sense of the present without a sense of the past, for Heidegger there is no present without a sense of the future. That is why Heidegger describes the ecstatic-horizonal future as "the datable and significant 'then,'" in contrast to the sheer or shorn Now that "has not yet come along but is only coming along."[48] Heidegger contrasts the ordinary metaphor of the Now as "pregnant" with the not-yet-Now to the more basic movement whereby "the Present arises from the future in the primordial ecstatical unity of the temporalizing of temporality."[49]

Insofar as the past and the future are built into the present, the metaphysical problem about the lack of a phenomenal difference between Nows that are *no longer* and Nows that are *not yet* does not arise. The directionality of temporality (the sense that the present arises out of the future, as opposed to coming from the past) is a condition for any temporal experience and for any experience whatsoever. Of course, Heidegger would not use the term "experience" because of its suggestion of subjectivity. Instead, one should talk about intelligibility, or the conditions for things showing up in the world and mattering to us.

With this account of temporality, the difference between Husserlians and Heideggerians becomes more clearly the difference between those Husserlians who hold that prereflective self-awareness is the key to the unity of consciousness and those Heideggerians who maintain that the unity of subjectivity is a function of our being-in-the-world and not just of our inner life. For Husserlians, the awareness that one is oneself having the experience is what accounts for the connectedness of experience. As we saw in the last chapter, Husserlians tend to think that subjectivity, or one's sense that one is having each experience that one has, is the key to the unity of experience. For Heideggerians, in contrast, it is not subjectivity but being in a *world* that is *intelligible* that accounts for the connectedness of experience. In contrast to the dualism of the traditional contrast between subjectivity and objectivity, world and intelligibility are more closely connected.

In more technical language, the difference is between Husserl's notion of intentionality and Heidegger's concept of transcendence. In *The Metaphysical Foundations of Logic* (1928), Heidegger explains that transcendence is not a movement from interior to exterior, as cognitive intentionality is often understood. The intentionality of the mind means that consciousness is always *about* something. Heidegger wants to show that although consciousness has the structure of intentionality, intentionality depends on transcendence, which is "the primordial constitution of the *subjectivity* of a subject."[50] Transcendence, or Being-in-the-world, is not simply one possible way of relating to beings. Instead, transcendence makes all relations to beings possible in the first place. Transcendence thus makes intentionality possible, and not the reverse.

Technicalities aside, the Heideggerian point can be made in either of two ways. One way is to say that the intelligibility of the world comes first, and the subject–object distinction is a distinction between different kinds of worldly experiences. Another way is to avoid the subject–object distinction and take the notion of Being-in-the-world seriously, especially the insistence not only on *Jemeinigkeit* or mineness, but also on the grid of intelligibility that makes up worldhood. *Jemeinigkeit* is the prereflective sense that Dasein has of itself, such that it can identify an experience as something that it is having. *Jemeinigkeit* is prior to reflective subjectivity. Similarly, the worldhood of the world is the way that the world presents itself, the way in which the whole is disclosed. Worldhood is prior to objectivity, and makes objectivity possible. No worldhood, no objects. Worldhood is not itself a specific content. Instead, it is that which makes it possible for content to appear as content, that is, as a feature of the world.

This reading of Heidegger is confirmed by his account of "world-formation" in *The Fundamental Concepts of Metaphysics*. There he explains his thesis that "man is world-forming" in contrast to an idealist way of understanding this phrase. The idealist will assume that Heidegger means by this thesis that "the world is nothing in itself but rather something formed by man, something subjective."[51] Heidegger distances himself from this subjectivism and suggests instead that world-formation is the ground without which the human being could not exist as such. More precisely, his thesis is that it is "the *Da-sein in* man" that is world-forming. The use of the hyphenated word "Da-sein" is a technical way of bringing out that only through the formation of the world could an individual show up for itself as an entity in that world. Subjectivity, like objectivity, is thus derived, not originary. He asks rhetorically, "How can man even come to a subjective conception of beings, unless beings are already manifest to him beforehand?"[52]

Reflective self-awareness enters the scene only when there is a breakdown in Dasein's way of encountering the world. A breakdown leads to reflective articulation and to the present-to-hand (*Vorhandenheit*). When a tool breaks, for instance, it becomes an object, that is, it is thematized as an explicit entity. But now that it is broken, it is no longer the tool, but only a piece of junk. The *purely* present-at-hand, where the object is viewed as it is "in itself," is not what the object is most primordially. Instead, the object is useless. Torn out of context, it is now merely in the way as a piece of scrap. For Heidegger the philosophical tradition has been making the mistake of starting with this decontextualized abstract understanding of the nature of things.

Heidegger thinks that the tradition is equally mistaken in its understanding of time. Instead of worrying about whether time is subjective or objective, or real or ideal, Heidegger wants to make these oppositions irrelevant by showing that temporality is prior to clock time. He does not thereby denigrate or deny the reality of clock time. On the contrary, he insists on the usefulness of time measurement. He also thinks, however, that by seeing that temporality is more primordial than clock time, we can become less lost in the lost Present, and more authentic in relation to our own finitude.

With this analysis of the existential significance of temporality and finitude, Heidegger could perhaps be perceived as going beyond the purely phenomenological concern with time-consciousness. Let us now return, therefore, to the phenomenological analysis of *Zeitbewusstsein*, particularly as it is developed in the work of Maurice Merleau-Ponty. The scene shifts, therefore, from 1927, when *Being and Time* appeared, to 1945, the date of the publication of Merleau-Ponty's most important book, *The Phenomenology of Perception*.

Merleau-Ponty on Temporal Idealism

In the *Phenomenology of Perception* Maurice Merleau-Ponty agrees with the foregoing discussion that the Now is an artificial way of construing the present. In his chapter entitled "Temporality," which in its central section represents an extended commentary on Heidegger's Kant book as well as an interpolation of Husserl's theory, Merleau-Ponty begins by taking issue with the standard ways of thinking of time both as a string of pearls, that is, as a series of instants, and as something that "flows" like a river. Even if one gives up trying to theorize time as an external physical process and moves toward viewing it as an internal conscious process, he thinks that it is wrong to suppose that time is a succession of nows. "We should," he says, "gain nothing by transferring into ourselves the time that belongs to things, if we repeated 'in consciousness' the mistake of defining it as a succession of instances of now."[53] So if it is a mistake to think of objective time as a sequence of nows, it is an even greater mistake to view temporality (as I use the term) that way as well.

Merleau-Ponty discusses the metaphysical view that objective reality by itself is a plenum, such that there is no room for time. On this view, time is not a "real process," and it is neither an "actual succession" nor a flowing substance.[54] In the plenum there is no time, because there can be no change and no events. Instead of thinking of temporality as a river or a string of pearls, he asks us to entertain the image of a fountain. He clearly has in mind a simple fountain such as is found in the Parisian public gardens with a single jet of water shooting up and falling back on itself. The fountain reinforces his account of temporality as an upsurge: "we are," he says, "the upsurge of time."[55] The fountain is at once an image of eternity and a sign of the constancy of the present. Merleau-Ponty in fact privileges the present: "Time exists for me," he asserts, "because I have a present."[56] In the broad sense of the present, which includes the horizons of the immediate past and future, he maintains that despite the fact that temporal modalities cannot be deduced from one another, "the present nevertheless enjoys a privilege because it is the zone in which being and consciousness coincide."[57] Insofar as this statement is not immediately clear, elucidating his case for prioritizing the present will require some explanation of other aspects of his conception of temporality. The danger of his view is that it comes close to the subjective idealism that is worrisome in Kant's account, and that Kant wants to avoid. In the elucidation that follows, this problem will be a continual concern.

Merleau-Ponty maintains that for there to be events, there must be someone to whom the events happen. Similarly, for there to be time, there must be an observer. The embodied observer supplies the reference point from which it first becomes possible to have change, and thus temporality. Merleau-Ponty does not think that time is a feature of objective reality in itself. If there were no subjectivity, there would be no time. He writes, "The past, therefore, *is* not past, nor the future future. It exists only when a subjectivity is there to disrupt the plenitude of being in itself, to adumbrate a perspective, and introduce non-being into it."[58] If this were the entire story, he would thus be a temporal idealist.

One objection that he would face is, what was the world like before there were people in it? Another objection concerns the phenomenology of temporal social experience, for he even says that time arises from "*my* relation to things."[59] Time is thus not first public, shared, and social, but is keyed to my particular subjectivity. An issue will be, then, how Merleau-Ponty can account for shared, public time if he starts from the assumption that time arises for each of us individually. If each of us has a different time, where does our sense of being *in* time come from? He will have to accommodate this view that time does not exist in objective reality, but arises from individual experience, with the phenomenology of time's passing, and the inexorability of this passing. Insofar as clock time is not regulated by individual time, but individual time is regulated by clock time, Merleau-Ponty may appear to have reversed the phenomenology of time.

Idealism is not the entire story, however, because Merleau-Ponty is trying to adapt Heidegger's interpretations of Kant that we saw in our earlier discussion of Heidegger's book, *Kant and the Problem of Metaphysics*. In particular, like Heidegger, Merleau-Ponty maintains that the very distinction between subject and object is derived from the more primordial structure of Being-in-the-world. Furthermore, because temporality temporalizes, we will have to rethink the nature and relation of both subjectivity and temporality. This rethinking will enable Merleau-Ponty to generate an answer to the above objections to his account, as I will show shortly.

For Merleau-Ponty, it is the metaphysical tradition that is at fault for making time incomprehensible. He remarks that it is often said that "the future is not yet, the past is no longer, while the present, strictly speaking, is infinitesimal, so that time collapses."[60] The string of pearls metaphor thus deconstructs itself. For if the past and the future do not exist, then the Now disappears because a present without a future or a past is not a present at all.

The only possible conclusion from this deconstruction is that the present is not a self-contained moment, a pearl in and of itself. For Merleau-Ponty, the past and the future are not separate existents so much as components of the present that make the present what it is. In phenomenology something that makes something else what it is can be said to "constitute" that phenomenon. In that sense the present is constituted by the past and the future. If that is the case, however, Merleau-Ponty's attempt to avoid deconstruction is still in trouble. For then the present would be nothing but a gap between two relata that themselves do not exist. In Merleau-Ponty's terms, the present would be only a "trace" left by two other traces, the past and the future, which are themselves nonexistent. Like the impression left after an erasure, a trace is not so much the presence but more the absence of a determinate mark.

Consciousness manages not to be imprisoned in this pure present without any transcendence toward the future or the past because consciousness "unfolds or constitutes time."[61] Neither is time a datum within consciousness; nor is time ever completely constituted, because consciousness is never completely constituted. A completely constituted consciousness would be Husserl's absolute consciousness, which is above time, and thus, an eternity.

For Merleau-Ponty, in contrast, what follows from the incomplete constitution of consciousness is that consciousness is always in the present. This is the fuller explanation, then, of why Merleau-Ponty privileges the present:

Time exists for me because I have a present. It is by coming into the present that a moment of time acquires that indestructible individuality, that "once and for all" quality, which subsequently enables it to make its way through time and produce in us the illusion of eternity.[62]

Presence is neither subjective nor objective because it is prior to the very distinction of subject and object, a distinction that is produced only by abstraction from presence.[63] This abstraction loses sight of the fact that temporality is not given as "an object of our knowledge, but as a dimension of our being."[64] Consciousness deploys itself in a "*field* of presence" in which the past and the future figure and from out of which the subject–object distinction arises. The idea of the field allows Merleau-Ponty to capture the phenomenological sense in which the present allows room for maneuver.

The problem with the view that we are only ever in the present is that it is then unclear how we could ever acquire the concepts of past and future.[65] At this point Merleau-Ponty shows us Husserl's diagram of time as a series of moments that are not isolated like pearls on a string, but are interconnected in such a way that "with the arrival of every moment, its predecessor undergoes a change."[66] (See figure 3.1.) The moment that moves into a past is retained by the new moment that takes its place. Then, "when a third moment arrives, the second undergoes a new modification; from being a retention it becomes a retention of retention, and the layer of time between it and me thickens."[67]

Merleau-Ponty raises the question whether Husserl's diagram brings us any closer to a clear understanding of temporality, or whether it simply restates the problem.[68] What Merleau-Ponty sees as the problem is that on Husserl's account there are still countable, distinct moments. Thus, Husserl's view still relies to some extent on the traditional picture of time as a series of instants. This can be seen when Merleau-Ponty goes on to say that "What is given to me is A transparently visible through A-prime, then the two through A-double-prime, and so on, *as I see a pebble through the mass of water which moves over it.*"[69] This picture of the pebble seen through the stream is, as Merleau-Ponty is aware, highly misleading. Unlike the pebble, which remains in sight while the water moves over it, moment A is gone, and is superseded by A-prime. A Bergsonian could well object, as we will see in the next chapter, that Husserl's diagram spatializes time into a series of moments, however interlaced they are, in the very act of trying to overcome the spatialization of time. Instead of seeing temporality as "a multiplicity of linked phenomena," Merleau-Ponty prefers to think that temporality is "one single phenomenon of lapse."[70] In fact, the lapse of time is taken as a proof of the oneness of temporality: "What does not elapse in time is the lapse of time itself," he writes.[71] The oneness (in my use of the term, as distinguished from the unity over time) of temporality is what Merleau-Ponty wishes to emphasize: "Time is the one single movement appropriate to itself in all its parts."[72] Or again, "there is one single time which is self-confirmatory" and "I am myself time, a time which 'abides' and does not 'flow' or 'change.' "[73]

The better analogy that Merleau-Ponty adapts from Husserl is the perception of a three-dimensional object, such as a box. When I see a box, I automatically presuppose, and even "perceive," the hidden sides and corners of the box.[74] Husserl and Merleau-Ponty maintain that if one did not in some sense perceive the hidden (or "absent") corners of the box, one would perceive simply a complex two-dimensional shape for which we do not even have a name. I will call it an intersection of planes, noting that even the shape of intersecting planes has another side that I cannot see. In any case, the moral of this story for present purposes is that just as the box would not be perceived as a box if one perceived the hidden corners as not being there, so the present could not be experienced as a presence without the adumbrations of the past and the future.

These protentions and retentions are intended to rule out the view of temporality as a series of Nows. The metaphor of a string of pearls, or a series of instants placed end to end in a line, is supposedly what Husserl should be avoiding, or what an interpreter should avoid attributing to Husserl. Instead, as Merleau-Ponty interprets Husserl, temporality is not a line, but a "network of intentionalities."[75] Merleau-Ponty objects to the metaphysical picture of objective time as a series of fixed positions at which we gaze. He also finds fault with time construed as a series of snapshots such that viewing them quickly enough will give us the cinematic experience of motion. His larger view is that "I do not so much perceive objects as reckon with an environment."[76] He thus thinks that he can avoid idealism by asserting that Husserl's protentions and retentions "do not run from a central I, but from my perceptual field itself."[77] That is to say, contrary to subjective idealism, temporality is not imposed by the mind onto the world, but grows out of the perceptual field in which it first becomes possible to distinguish mind and world.

This argument allows him to add Heideggerian elements to the Husserlian notions of retention and protention, thereby "enriching" the Husserlian diagram. The future is described as "a brooding presence moving to meet one, like a storm on the horizon."[78] In fact, Merleau-Ponty's privileging of the present leads him to oppose Heidegger's prioritization of the future. He thinks that Heidegger's privileging of the future is impossible for Heidegger to accept both on his own grounds and in view of the phenomenology involved. If in our everyday life we are always centered in the present, as Merleau-Ponty understands the temporal phenomenology, and if the everyday present is inauthentic, as Heidegger believes, then how can we be anything other than inauthentic? Furthermore, Merleau-Ponty's future is different from Heidegger's future. Whereas for Heidegger we project our possibilities *forward* into the future, for Merleau-Ponty the future is really a form of retrospection insofar as I generate my views of the future based on past experience. Merleau-Ponty thus speaks of *anticipatory retrospection*, which projects the future backward into the past in the very act of looking forward to what is coming next.

On my reading, Merleau-Ponty is further enriching Husserl when Merleau-Ponty says that temporality is not a series of objective positions through which we pass, but a "mobile setting" that moves in relation to us. Merleau-Ponty describes temporality as a "bursting forth" or dehiscence, and he then remarks, "Hence time, in our primordial experience of it, is not for us a system of objective positions, through which we pass, but a mobile setting which moves away from us, like the landscape seen through a railway carriage window."[79] As I interpret him, he is saying that temporality has a horizonal character, much like spatiality. Thus, just as the hill in the distance remains relatively still while the trees close to the tracks whiz by in a blur, so some temporal features will remain relatively stable while others will rush past.

This image clarifies what he means when he says that temporality is a function of the perceptual field and is not imposed by a Kantian central I. But when he says that temporality is a performative, and that it is *I* who performs the ecstasis, he realizes that subjectivism lurks close at hand.[80] So he is quick to add the Heideggerian argument about temporality being the core of subjectivity: "We are not saying that time is *for* someone. . . . we are saying that time *is* someone. . . . We must understand time as the subject and the subject as time."[81] He then adapts Heidegger's criticism of Kantian faculty psychology to his own purposes. The criticism is that it is impossible to understand how a transcendental ego could ever become aware of itself in time since there is no content of which one could be aware. Noting that Heidegger attributes self-affection to temporality, Merleau-Ponty argues,

If, however, the subject is identified with temporality, then self-positing ceases to be a contradiction, because it exactly expresses the essence of living time. *Time is "the affecting of self by self"*: what exerts the effect is time as a *thrust* and a passing towards a future; what is affected is time as an unfolded series of presents; the affecting agent and affected recipient are one, because the thrust of time is nothing but the transition from one present to another. This *ek-stase*, this projection of an indivisible power into an outcome which is already present to it, is subjectivity.[82]

This is an explanation, then, of how temporality and self-consciousness are connected. Because temporality is subjectivity, and subjectivity is temporality, temporality also has the capacity to be aware of itself. This may seem a strange thing to say, but we have seen it already in Heidegger. Indeed, Merleau-Ponty cites Heidegger's Kant book in asserting that not only is temporality modeled on subjectivity but also that subjectivity is modeled on temporality. In notes from 1959 that he did not publish, Merleau-Ponty appears to have learned from Husserl as well "that it is not I who constitutes time, that it constitutes itself, that it is a *Selbster-scheinung*."[83] Paraphrasing Heidegger, Merleau-Ponty says in the *Phenomenology of Perception*, "It is the essence of time to be not only actual time, or time which flows, but also time which is aware of itself, for the explosion or dehiscence of the present toward a future is the archetype of the *relationship of self to self*, and it shows up [as] an interiority or ipseity."[84]

What should be remembered here is that Merleau-Ponty wants to illustrate the underlying connection to the world that Heidegger calls transcendence. Merleau-Ponty's success can be questioned, however, insofar as his statements still sound subjective. For instance, when he says, "we are the upsurge of time," the reference to the "we" connotes the mind-dependence of time.[85] Similarly, the claim that there can be no directionality or movement in the world in itself without the subjectivity of the perception of an observer reintroduces the subject at a primordial level of the account.[86] He grants that the subject requires world; but at the same time he asserts, "the world remains 'subjective' since its texture and articulations are indicated by the subject's movement of transcendence."[87]

These statements invite the objections that I mentioned earlier. The first of these takes issue with Merleau-Ponty's argument that there could not be a world without human beings. Merleau-Ponty himself raises and responds to this objection as follows.

What, in fact, do we mean when we say that there is no world without a being in the world? Not indeed that the world is constituted by consciousness, but on the contrary that *consciousness always finds itself already at work in the world.* What is true, taking one thing with another, is that there is a nature, which is not that of the sciences, but that which perception presents to me, and that even the light of consciousness is, as Heidegger says, *lumen naturale,* given to itself.[88]

In other words, nature as described by the natural sciences is reduced to laws and quantifications that often have little import for my everyday perceptual experiences. We should not look to physics, then, for an account of our temporality and our experience of Being-in-the-world. Where there is world there is consciousness already at work, which is not to say that first there is consciousness, and only then is there a world.

More pertinent to a discussion of temporality is Merleau-Ponty's response to our question about the intersubjectivity of temporality. The issue is, if temporality is so much a function of *my* perceptions and *my* performances, how do we explain its phenomenal oneness and unity? His rejoinder is worth citing in its entirety, especially insofar as it reflects a similar response by Husserl.

It is true that the other person will never exist for us as we exist ourselves; he is always a lesser figure, and we never feel in him as we do in ourselves the thrust of time-creation. *But two temporalities are not mutually exclusive as are two consciousnesses, because each one arrives at self-knowledge only by projecting itself into the present where both can be joined together.* As my living present opens upon a past which I nevertheless am no longer living through, and on a future which I do not yet live, and perhaps never shall, *it can also open on to temporalities outside my living experience and acquire a social horizon, with the result that my world is expanded to the dimensions of that collective history which my private existence takes up and carries forward.* The solution of all problems of transcendence is to be sought in the thickness of the pre-objective present, in which we find our bodily being, our *social being*, and the pre-existence of the world, that is, the starting point of "explanations," in so far as they are legitimate—and at the same time the basis of our freedom.[89]

The argument is that whereas the first-person access that consciousness has to itself is private rather than public, the same could not be said of temporality. Temporalities are more readily merged into a single time. The difficulty here is that the phenomenology is reversed. Instead of finding ourselves *already in* a present, we seem to have to *project* the present. This account thus appears to put the cart before the horse and to lose sight of the unavoidability of the present.

However the phenomenology shakes out, on Merleau-Ponty's picture the present is the overlap of past and future, and this continuous overlapping is the passing of time. Of course, instant A and instant B are not indistinguishable, for otherwise there would be no time. The subjective standpoint distinguishes the events that order temporal instants into before and after. Instead of emptying the present out by saying that it is merely the gap between past and future, Merleau-Ponty fills up the present in a Bergsonian manner by saying that the present is "one single time" in which the whole past and the entire future are present.[90] The passage of time is thus not something that I passively observe, but instead, something that

I effect. "I am myself time," we saw him assert.[91] This point allows him to retain the river metaphor, but not for its flow or its successive multiplicity. Instead, the river serves as an image for the unity or permanence of time. This metaphor also supports his privileging of the present as "the zone in which being and consciousness coincide."[92] The next question to take up is, then, whether this privileging of the present makes Merleau-Ponty susceptible to Derrida's criticism of phenomenology for falling back into the metaphysics of presence. I shall therefore need to explain in more detail exactly how this criticism works.

Derrida's Critique of the Metaphysics of Presence

Heidegger and Derrida both offer a critique of how the present is used in the history of metaphysics. In Derrida in particular, this critique takes the form of a challenge to "the metaphysics of presence," which can show up as "logo-centrism," "phono-centrism," or "ethno-centrism." He attacks not only the metaphysical bias that favors the present in the temporal sense, but more thoroughly, he deconstructs the prejudice of the entire Western tradition that makes presence the exclusive paradigm of philosophy. Challenging the tradition at its deepest level is what makes Derrida important now and in the future.

What, then, is the metaphysics of presence? Can it be avoided, or is it built into what is now called "thinking," or "theory," or "philosophy"? Presence is not simply the direct connection of a subject and an object, but it is tied to subjectivity as such. Speech is at once an example of this presence and the paradigm of it (a paradigm being the most privileged example or case of that which it illustrates). Philosophers traditionally privilege the present moment when a speaker utters a sentence that directly refers to the world. The intuition is that the speaker is less likely to be wrong than when the speaker is referring to a state of affairs that is not directly perceived. In the present moment the appearance and the appearing are apparently one and the same. Derrida, in contrast, thinks that there is always a gap between appearance and reality that comes before the moment of their coinciding. Without a prior difference the moment of identity would not be possible. He calls this difference the "trace." He does not acknowledge Merleau-Ponty as a source of this notion of the trace. Merleau-Ponty had said in a Bergsonian moment: "This table bears traces of my past life, for I have carved my initials on it and spilt ink on it. But these traces in themselves do not refer to the past: they are present."[93] Derrida may have felt that Merleau-Ponty's use was still mired in the metaphysics of presence, given this Bergsonian privileging of the present. Instead of Merleau-Ponty, Derrida identifies Emmanuel Levinas as the source of his term "trace" when Derrida says that he is merging Levinas's term with a Heideggerian intention of destroying ontology. *Of Grammatology* specifies the trace by remarking, "The unheard difference between the appearing and the appearance [*l'apparaissant et l'apparaître*] (between the 'world' and 'lived experience') is the condition of all other differences, of all other traces, and *it is already a trace*."[94] Although the trace might seem to be the most metaphysical concept of all, he thinks that the trace cannot be grasped by metaphysics and thus puts us beyond metaphysics.

The question is, however, whether the incomprehensibility of the trace to the metaphysics of presence makes the concept of the trace unintelligible. After all, the conditions of intelligibility are configured *within* the metaphysics of presence and would not be applicable to a thought that was *outside* that tradition. Derrida recognizes that a metaphysics that starts from a conception of the plenitude of presence based on the paradigm of speech will not be able to make sense of his notion of the trace. At the same time, the notion of trace cannot entirely escape this metaphysics of presence. Derrida grants the "ambiguity" of the concept of the trace, which is both the absence of a presence and the presence of an absence. The fact that ambiguity is itself a notion that is tied to the metaphysics of presence means, he is aware, that the trace "requires the logic of presence, even when it begins to disobey that logic."[95] Derrida's metaphilosophical intention in using notions such as the trace is to generate an entirely different kind of philosophy from that which starts from the paradigm of presence. Instead, his thought is that an alternative way of thinking will evolve from starting with the paradigm of writing and explaining how making sense is possible even in the absence of a speaker or author.

There is, of course, a difference between ontic or ordinary instances of traces and the more philosophical concept of trace that Derrida is developing. The trace in the ordinary, ontic sense is nicely illustrated in a deleted scene from the film on Derrida.[96] Derrida is explaining in the voice-over that the trace of a person, for example, persists for a brief period after the person has died. Anticipating his approaching demise, perhaps, the camera in this scene does not show him but only traces of him: his watch lying on his bureau, his pipes on their stand, photos of him in his youth, his handwritten shelf labels for the books of Heidegger, Bourdieu, and others in his library. In contrast, the trace in the more philo- sophical or quasi-ontological sense is the basic unit of grammatol- ogy based on the paradigm of writing, much as the sign (construed as the unity of signifier and the signified) is the basic unit of lin- guistics based on the paradigm of speech. The grammatological approach displaces traditional concepts such as subjectivity and consciousness, and thus deconstructs the philosophies that privi- lege them.

Derrida's critique of Husserl in *Speech and Phenomenon* identi- fies the source of the illusion of presence as the voice. Hearing oneself speak (*s'entendre-parler*) is the pure self-affection that generates one's sense of oneself as being a subject who is produc- ing the speech. Subjectivity is under the illusion that it is the source of the meaning of what it says, when in fact that meaning is what first makes possible saying what one wishes to say. As a corollary, the sense that one has of oneself as being a subject that constitutes

one's own experience is mistaken. Instead, subjectivity is constituted by the phenomenon of hearing oneself speak. Even those who
are deaf or mute must mime speaking. "No consciousness is possible without the voice," we are told, and "the voice *is* consciousness."[97] Not only does the voice generate the sense of oneself as a
subject, it also makes it possible for the world to appear as independent of us: "This auto-affection is no doubt the possibility for
what is called *subjectivity* or the *for-itself*, but without it, no world
as such would appear."[98]

Furthermore, this auto-affection generates the illusion of time as
a *movement* from one "living present" to another, and it covers up
the metaphorical character of the word "time." For Derrida, the
living present is not really fully self-present, but instead is "always
already a trace."[99] The history of metaphysics has tried to cover up
the illusion we have that the self of the living present is primordial.
What Derrida suggests is that temporal difference is what generates
the sense of the living present as being self-same, rather than the
other way around. In other words, temporalization makes possible
the conceptual distinctions between subject and world, inside and
outside, existent and nonexistent, constituted and constituting, and
even space and time. Whereas the metaphysics of presence presupposes these distinctions and tries to explain experience in terms of
them, Derrida shares Heidegger's sense that thinking about temporality requires an explanation of how these distinctions emerge
from the more primordial activity of temporalization, or temporality temporalizing.

Traditional phenomenology, which assumes the self-presence of
subjectivity to itself, should be one of the metaphysical philosophies that drop away in the face of Derrida's hermeneutical emphasis on writing and interpretation. "Writing," says Derrida in *Of
Grammatology*, "can never be thought under the category of the
subject."[100] Just as writing can make sense in the absence of either
the author or the world as it was at the time of writing, so "the
original absence of the subject of writing is also the absence of the

thing or the referent."[101] The temporality of writing is different, therefore, from the temporality of speech. Whereas the paradigm of speech privileges the present, the temporality of writing is more revelatory of the past and future. In fact, in contrast to the eternal atemporality of the present, writing first makes the perception of temporality possible. This "writing of difference, this fabric of the trace," says Derrida in *Of Grammatology*, is the "*origin* of the experience of space and time" and it "permits the difference between space and time to be articulated, to appear as such, in the unity of an experience."[102]

This claim that writing is the "origin" (in the temporal sense of the more primordial as well as the logical sense of a more basic presupposition) of the experience of space and time may strike contemporary ears as being equally as metaphysical as Kant's claim that space and time are mind dependent. Indeed, Derrida appears to be tempted by the allure of transcendental philosophy, with its desire to come up with the most basic categories of experience that explain all the other features of experience. When he says that the trace makes all sense-making possible or when he asserts that the trace "is the condition of all other differences"[103] as well as of "the constitution of subjectivity" itself,[104] he seems to be trying to be more Kantian than Kant himself.

At the same time, however, he tries to undercut the very possibility of transcendental philosophy by denying that there are primordial categories. Thus, just when he seems to be falling into the project of transcendental philosophy, he insists immediately that there is no such project: "*The trace is in fact the absolute origin of sense in general. Which amounts to saying once again that there is no absolute origin of sense in general.*"[105] In other words, unlike metaphysics, which thinks of its basic concepts as self-contained units of meaning, Derrida's concept of trace is not such a unit. There are no such units but only contrastive relations in a system of differences. These differences are both spatial and temporal. Spatial relations are said to *differ* whereas temporal relations are *deferred*. He points to Freud's notion of the deferred effect, or *Nachträglichkeit*, as an example of a temporality that disrupts the usual conceptualization of time as involving the moments of present, past, and future. Freud's discovery of a present that is not immediately prior to the next one but considerably anterior to it suggests a very different temporality than is presupposed by "a phenomenology of consciousness or of presence" or by "the metaphysical concept of time in general."[106] Derrida's deconstruction of presence thus is carried out through a deconstruction of consciousness, particularly of Husserl's notion of "internal time-consciousness," although it could also be aimed at Merleau-Ponty. His analysis calls into question standard assumptions about what we call "time, now, anterior present, delay, etc."[107] We can no longer accept Kant's and Husserl's Newtonian view of time as linear succession, homogeneity, or "consecutivity," that is, unified, uninterrupted unfolding.[108]

The question is, where does deconstruction leave us in relation to our understanding of temporality, or time's passing? Derrida's analysis would not have been possible without the doubts raised about consciousness by Friedrich Nietzsche. Nietzsche maintains that consciousness cannot be trusted to know its own functioning. If that is true, then Husserl's method of phenomenological description must be replaced with a genealogical method that digs more deeply into underlying motivations and structures.

So let me now turn to Nietzsche's analysis of temporality.

Nietzsche and Deleuze on Eternal Recurrence

Nietzsche might seem to be an uncommon figure to include in a history of *phenomenology*. His pronouncements may seem to be more metaphysical and ethical than phenomenological, and he does not explicitly distinguish between time and temporality. The reason for including him in this chapter, however, is that he is trying to *change* our phenomenological sense of time. He is trying to get us to give up the sense of time as culminating in some remote eschatological future and instead come back to a more immediate focus on the need for action in the present. He is also trying to free us from the burden of the past, which weighs heavily on us. On Nietzsche's account, the will needs to be liberated from its nostalgia for the past and its sense of helplessness in the face of its inability to change the past. We have to survive the learning process whereby we come to see that the past does not justify the present, if only because nothing does. The Great Noon at which Zarathustra finally arrives after going through the midnight of overcoming suggests that living in the present without nostalgia for justification of the past or hope for redemption in the future should cause us joy rather than despair. This joy or *Heiterkeit* is to be secured through the doctrine of eternal recurrence. In the light of these considerations, then, Nietzsche deserves a place in this discussion.

According to *Ecce Homo*, Nietzsche thought that his idea of the eternal return was his greatest insight and that it was also Zarathustra's most fundamental idea. What is not clear is exactly what he thought that he saw in this insight. Questions that commentators often raise include the following:

1. What is it that recurs?

2. Is the account of time cyclical or linear?

3. Can we know whether eternal recurrence occurs? If we cannot know the truth of the hypothesis, then how does that change our attitude toward the cosmological claim?

4. Do our lives become better through this conjecture about time?

I will give my own answers to these questions sequentially before adding two further questions of my own.

First, what is it that recurs? Here the texts do not settle the question, and there is room for philosophical reflection. Would the repetition of every single little detail be required? The nausea with which the initial announcement of the doctrine is received suggests this interpretation. In §341 of *The Gay Science*, "The Heaviest Weight," Nietzsche describes the idea for the first time as follows:

What if some day or night a demon were to steal into your loneliest loneliness and say to you: "This life as you now live it and have lived it you will have to live once again and innumerable times again; and there will be nothing new in it, but every pain and every joy and every thought and sigh and everything unspeakably small or great in your life must return to you, all in the same succession and sequence—even this spider and this moonlight between the trees, and even this moment and I myself."[109]

The thought of the eternal recurrence of the smallest details is thus initially proposed as leading to a general nausea with one's existence. This nausea is caused by the past as well as by our lack of power to change it. The repetition of general patterns, however, is also often thought to be sufficient to generate nausea. Although I prefer the interpretation whereby what recurs is every detail, there are notes in Nietzsche supporting the claim that the repetition of patterns is all that has to recur. Nietzsche talks about, for instance, "the absolute necessity of *similar* events occurring in the course of one world, as in all others."[110] "Similar" does not mean the same, so the repetition of patterns would be all that has to occur.

Gilles Deleuze tackles the question differently. What he claims is that

It is not the same that comes back, since the coming back is the original form of the same, which is said only of the diverse, the multiple, becoming. The same doesn't come back; only coming back is the same in what becomes.[111]

For Deleuze difference is prior to sameness, so anything could only ever recur as different. Deleuze can thus say, "Nietzsche's secret is that *the eternal return is selective*."[112] The doctrine allows for the separation of active from reactive forces and the selection of the former. This interpretation allows for Deleuze's Nietzsche to be taken as denying that every specific event must recur.

Independent of the issue of whether what repeats is the same or the different, there are two problems with any interpretation, including Deleuze's, that holds that what recurs is the repetition of *patterns* rather than *specific events*. First, patterns are not as regrettable as specific instances of such patterns. One might very well not want to repeat a particular action over and over again, even while granting that the pattern is very likely to recur. To take an example, one might very well not want to live again and again having said a specific thing that had better been left unsaid. But the general pattern of regretting things that one has said is such a common part of human psychology such that one could live with that feature repeating itself over and over. Second, the doctrine is potentially vacuous if all it means is that general patterns recur. All events are alike in some basic respects, and if these basic respects are what recur, then any event could substitute for any other event.

The second issue I raised above concerns whether time in the doctrine of eternal return is cyclic. Nietzsche is often described as if he were attempting to substitute a cyclical conception of time for a linear one, such as is espoused by Christianity. Elizabeth Grosz points out, however, that the doctrine of eternal return posits time as an endless infinity that is neither cyclic nor linear. Grosz, who has taken Deleuze's reading of Nietzsche to heart, suggests that Nietzsche is not collapsing space and time into the space-time of modern physics. "What recycles," she maintains, "is never time itself but what exists in time: things, processes, events, formations, constellations, in short, matter in all its permutations."[113] There is recurrence just because time is infinite and matter is finite. Time and matter are conceptually separate, and matter is conserved, but time "squanders itself without loss."[114] The infinity of time is required for there to be enough time for all the various combinations of matter to be able to recur.

The third question concerns whether we have to believe that the cosmological doctrine of eternal recurrence is true or at least possible. Since a proof of eternal recurrence is unlikely, will the very thought of eternal recurrence move us to the point of transforming our lives? Whether Nietzsche himself actually believed the cosmological thesis, and for how long, is questionable. The doctrine is announced by a demon in one place, for instance, and in a drunken song in another. Zarathustra's animals mouth it, and Zarathustra himself is a fictional character. Some formulations do not appear in print, but only in posthumous notes. Does it matter, then, whether Nietzsche espoused the doctrine or whether we adopt it for ourselves? Nietzsche scholars have suggested that eternal recurrence need not be true, only possible. If that is right, then the fourth question I raised above can now be answered, for even a possible thought of this magnitude—think of the effects of the thought of eternal damnation—can move or transform our lives. But then, how would we know that the doctrine is even possible, since in principle one cycle can contain no evidence of any previous occurrences of it?

Because of these difficulties, there are advantages to thinking of the doctrine of eternal return as a *thought experiment*. A thought experiment need not be believed to be true or even possible to function effectively as a test for our ability to affirm life. Thought experiments are common in philosophy today, and we are familiar with the way they reveal conceptual entailments as well as the way in which they oversimplify complex cases. Treating the issues as thought experiments takes the burden off any inability to believe in the hypothesis. At the same time, however, the advantage is also a disadvantage. The disadvantage of seeing eternal return as a thought experiment is precisely that the strength of belief posited in a thought experiment is minimal. Take the case of eternal damnation. If eternal damnation had been thought to be merely a thought experiment, it could not have been a sufficiently strong belief to frighten people for centuries.

Recognizing the advantages and disadvantages of thought experiments leads me to add two further questions to the list:

5. If eternal recurrence is a thought experiment, what does its success or failure establish about our attitude toward time?

6. How does the hypothesis of eternal recurrence relate to temporality, if at all?

My reason for adding this last question concerns the large issue whether cosmological claims about time make any difference to our experience of temporality in our own lives.

In response to the fifth question, let me start by clarifying the question and asking, what are examples of having an "attitude toward time"? Often the expression "having an attitude" suggests a confrontational stance, and that is no less true in this instance. "*Ressentiment*" can be directed against time, and the projection of a timeless eternity may well be a case of revenge against time, given our inability to escape its march ever onward. If the thought experiment of eternal return is to naturalize human beings, then it should have an effect on this *ressentiment*. The goal should be to get rid of the idea of goals, of purposes, of a *telos* of the world and of the human. In *The Gay Science* §109 Nietzsche says, "Once you know that there are no purposes, you also know that there is no accident."[115]

Nietzsche thus rids us of both teleology and eschatology. Teleology is emptied of content, and eschatology goes out the window because there is nothing outside of time that could appear in time. Teleology without eschatology is empty, and eschatology without teleology is blind. Without them, however, how does the conjecture of eternal recurrence make our lives better, and in particular, how does it generate a more positive sense of our temporality and a better attitude about time? To deal with these issues, Deleuze reformulates the language to produce the Nietzschean equivalent of the Kantian categorical imperative: "*Whatever you will, will it in such a way that you also will its eternal return.*"[116] As in Kant's test of personal maxims to see whether they could be universalized into universal principles that everyone could rationally will, this formulation is thus a thought experiment to see whether, independent of the truth or falsity of the cosmological doctrine, the very thought of eternal return is sufficient to guide action. The thought experiment of the Nietzschean imperative is selective, on Deleuze's account, because it helps to select some forces as active and to deselect reactive forces. By consciously selecting active over reactive forces, we can change our relation to the past from a passive to an active one. We thereby live up to Zarathustra's secular acceptance of the past: "To redeem the past and to transform every 'It was' into an 'I wanted it thus!'—that alone do I call redemption."[117] The creative will takes an unresentful relation to time by adopting the attitude of reconciliation with time through willing backward. If we cannot transcend time, we can at least learn to accept it without *ressentiment*. In *Ecce Homo* Nietzsche calls this attitude *amor fati*, love of fate: "My formula for greatness in a human being is *amor fati*: that one wants nothing to be other than it is, not in the future, not in the past, not in all eternity. Not merely to endure that which happens of necessity, still less to dissemble it—all idealism is untruthfulness in the face of necessity—but to *love* it."[118]

Through *amor fati*, therefore, the character of temporality is changed. We learn to affirm not only the past but also the irreversibility of temporality.[119] Insofar as one cannot undo what has been done, there is no point to lamenting the past. Yesterday's pop culture would have said that we have to learn "to take responsibility" for past deeds, whereas the current expression is that we have to "deal with it." Both of these expressions overlook the complexity of transforming *ressentiment* into *amor fati*. But they do suggest the move away from the negative relation of being "in denial" or "reactive," and the transformation of temporality into an active or creative or "constructive" attitude.

In this sense, Nietzsche's transformation of temporality bears a certain resemblance to Heidegger's. Both take a negative phenomenon and turn it into something positive. Unlike pop culture, they do not simply invert the relation. Neither Nietzsche nor Heidegger is discussing attitudes that are merely "inner." Instead, they are calling for a change in the kinds of actions that are to result. To affirm a more positive temporality is not simply to have a more "positive outlook" on life. Affirmation is not simply an inner attitude. Instead, it involves more outward-directed action. Through action rather than attitudes, the creation of positive situations becomes possible as one takes constructive measures in the world over time. What makes situations positive is that they enable rather than discourage experiments that involve not simply an individual's inner self (as Kierkegaard implies is the case with Abraham), but other beings as well.[120] Because the world is always already social, and because our actions impinge on others, positive action means increasing the possibilities for action, not simply for oneself, but for all agents.

To sum up, if the eternal return is a nauseating thought, the nausea is not due to the trivialities that will repeat themselves over and over. For the most part, we like trivialities and enjoy them. Dying people often list trivial things and events as what they are saddest about losing.[121] What makes the thought experiment of eternal return difficult to bear is the profound boredom that results from it. I use this term in Heidegger's sense because it is one side of the coin of eternal recurrence. Like eternal recurrence, profound boredom is boredom with time itself, and particularly with the past. The other side of the coin includes our joy at being liberated from the burden of the future. On my reading, the affirmation of the present is the key outcome of Nietzsche's thought experiment. Alexander Nehamas is right to say that the eternal return is the final test of the integrative process that aims at a coherent, but never completed whole, with no constant elements or even a determinate number of them.[122] If the eternal return implies that the stories of our lives are always subject to reinterpretation, and if Nietzsche's thought experiment leads us successfully to see that point, then Zarathustra can say to death: "Was *that* life? . . . Well then! Once more!"[123]

This conclusion should be uplifting, but there is a remaining issue that must be confronted. Beginning readers of Nietzsche often fail to note the irony of producing a metaphysical doctrine to establish a naturalistic account of the human. The purpose of the cosmological hypothesis is to get us to see that there is no outside to time, and thus to explain our lives as part of the natural world without any need for or means of transcending it. A purely naturalistic theory would have no need, however, for a cosmology that had no causal role in this universe. Insofar as Nietzsche did not envision alternative universes with different causal laws, but only this one endlessly spinning away, the question is whether the story of one cycle and no more is not more naturalistic and believable than the story of an infinity of recurring cycles.

A much simpler hypothesis than the eternal return is the "one time only" story. In other words, the point about the reinterpretability of our lives is independent of the theory of eternal return. Explanatory parsimony by itself suggests the adoption of the hypothesis of one life only. Although the doctrine of eternal return might have provided some countervailing pressures against theological doctrines, once God dies and theology loses its purchase, eternal return becomes simply another unnecessary metabelief. Unless some nontheological advantages of eternal return are proposed, therefore, I suggest that eternal return is itself a hypothesis for which we no longer have any need.

Reflections

Let me now summarize the main points that have been asserted by these philosophers and add my own hermeneutical observations. Looking over these theories of the present, there is an almost universal dismissal of the view of temporality as a series of Nows. In fact, this view is so often attacked that it is not clear whether anyone in modern philosophy actually holds it. All these views—including Hegel's critique of sense-certainty, Husserl's analysis of protention and retention, Heidegger's ecstases, James's notion of the dawning and fading of each moment of consciousness, Merleau-Ponty's notion of the identity of time and the self, Nietzsche's *amor fati*—want to build the past into the experienced Now, and some want to do as much for the future as well.

Given this consensus, a brief review of some points in this chapter may be timely. In Hegel the problem with the Now threw us back to the dependence of temporality on the observer. Sense-certainty tries to find its certainty in the objectivity of protocol statements with "Here" and "Now" built in. Hegel then provides two arguments to show the observer-dependence of temporality. First, the Now that we can point to is not the Now that currently obtains, but one that is always already in the past. Second, the Now is infinitely divisible into smaller and smaller units. The Now is thus not what sense-certainty thought it was, namely, something objective that we know through immediate access. Instead, what we learn is that if there were no observer, there would not be any earlier or later, any before or after. Sense-certainty's project of getting in touch with the world stumbles insofar as what is discovered is not objective but subjective.

Whereas Hegel's brief discussion is strictly critical, William James in his *Principles of Psychology* has made a constructive effort to describe lived time from the inside. James's reflections go beyond Hegel's by advocating that one give up the idea of the Now and recognize instead the prereflective unity of past, present, and future in self-awareness. James wants us to think about temporality not as a digital (on or off) phenomenon, but as an analog one (with warm-up and fade-out phases).

Heidegger joins this attempt to build duration into the Now. He could be compared with Bergson, except that he thinks Bergson's strategy of externalizing a qualitative time is misguided insofar as it sells the reality of objective time short.[124] Insofar as datability, significance, spannedness, and publicness are built into each Now, these features show how Nows can be distinguished from one another. For instance, datability makes it possible for before and after to be remarked, and also for temporal ordering to obtain, whether or not memory is involved.

Heidegger's theory insists on the priority of temporality over time. Time is reduced to ordinary clock time, which is not primordial. Lived temporality is what is primordial. Neither a staccato of pearls nor a fluvial flow, temporality is made up of action-oriented practices. These practices interpret the situation into relational ecstases of projecting a future that may never come onto a past that perhaps never was. Because temporality is always contextual, how long the present lasts will depend on the goal and the origin of the interpretive practice.

Heidegger did not finish his argument in *Being and Time*. On my reading, however, he was on his way to showing that temporality is hermeneutical, just as hermeneutics is always temporal. A crucial consequence of this hermeneutical account of temporality is that it becomes possible to say that in a certain sense (to be discussed in chapter 3) we can change the past. Changing our conception of what is happening now in the present can lead to changes in the account of what has happened in the past. This change of the present is tied to a moment of vision that sees the past and present in a different light by projecting a different future. Because the future is never fixed but is always the horizon of an interpretation (as will be discussed in chapter 4), it is eminently changeable and always up for grabs. Future truth-values are less constrained than those of the past and to some extent even of the present. They are not entirely unconstrained, however, since a projection of a future depends on the coherence of connecting that future to a particular understanding of the past. Consistency is itself a constraint on the range of possible interpretations.

To conclude this summary, here are some hypotheses about phenomenological differences between time and temporality, to be explored further in the next two chapters.

• If time marches on inexorably, temporality is sometimes fast, sometimes slow.

• If time is the grid of scientific intelligibility, temporality is the grid of lived intelligibility.

• If time is one moment and then another and another, temporality involves an ordering of not only before and after, but also longer and shorter.

• If time goes tick, tick, tick, temporality goes tick-tock, tick-tock.[125] In other words, there are no inherent qualitative differentiations in clock time; they are gained only through temporality.

• Temporality involves retention, protention, and the primal impression. These are not the same as the past, present, and future, which are dimensions of time.

• The Now that we experience temporally is never the Now of time, since this Now is always already gone by.

• Temporality, unlike the time of the universe, is irreversible.

• There is nothing in the moment of time that tells whether it is past, present, or future. The order comes from temporality. For James, there is no sense of the present of temporality without a sense of the past. For Heidegger there is no sense of the present without a sense of the future.

• Two temporalities are not as irreconcilable as two consciousnesses.

This list could be continued, but for now it illustrates the kinds of claims that a phenomenology of temporality could generate. Once these and other such differences are noted, there is really not much point in Heidegger's worry about whether time or temporality is the more primordial. I take the point of Derrida's deconstruction of transcendental philosophy to be that the question of which is grounded in which does no real work. Furthermore, the attempt to answer such a question represents a dubious philosophical vestige of Kant's program of transcendental philosophy. If understanding is always interpretive, there is no understanding of interpretation that is not also an interpretation. Temporality is a basic feature of interpretations of the world. An interpretation of the world will always have a temporal dimension, and if that temporality is changed, the interpretation will change as well. The following chapters will discuss whether changing the present will also change either or both the past and the future. We cannot determine that change, but we can try to change things for the better. What would be the point, after all, of trying to change things for the worse?

3 Where Does the Time Go? On the Past

This chapter is concerned with the past, with memory, and with the conditions for memorialization. What is the past? The past is sometimes construed as the present frozen in a kind of stasis. Science fiction is thus able to imagine time travel as a return to a time and place where what happened is still happening, just as the present is happening now. The only condition on the past is that it is closed, unlike the present, which still opens into the future. The future is sometimes assumed to be structurally like the past, except perhaps that it is less frozen and fixed than the past.

But clearly this way of imagining time is mistaken and misleading. When an experience moves into the past, it is over and done with. There are no other universes where things keep happening as they did in our universe. But then we are faced with the problem that the past still structures the present. If the past moments are totally gone, how can they continue to have this living relation to the present? To answer that, we must ask philosophical questions about the past. Where does the past go? Can the past be changed? How can something like the past, which is presumably unreal, nevertheless determine, condition, or influence the present?[1]

The selection of philosophers in this section is thematically constrained by the topic of the malleability of the past and the explanation of memory. After reviewing how these issues have been focused in contemporary philosophy, with particular reference to the work of Ian Hacking, I go on to review the contributions of the German and the French traditions to the discussion of these topics. In particular, Husserl, Heidegger, and Gadamer are taken as representative of the phenomenological and hermeneutical traditions in their approach to the past. On the French side, Jean-Paul Sartre, Jacques Derrida, and Pierre Bourdieu offer different kinds of accounts of memory and of the conditions for memorialization. This chapter then turns to the tradition of Bergsonism for an account of duration that is directly at odds with the Husserlian account. Bergson emerges, however, as a markedly different philosopher when interpreted by Maurice Merleau-Ponty in comparison to the picture we get of him from Gilles Deleuze. The difference between the two Bergsons reflects, of course, the underlying methodological difference between phenomenology and the poststructuralist application of genealogy. This contrast is explained in the postscript on method.

Phenomenology of the Past

Can the past be changed? "We choose our past," as Hayden White sums up Sartre on history, "in the same way that we choose our future."[2] Can this be true? Or is it rather the case that the past is over and done with? Is the meaning of the past fixed, such that the deeds that are done make us who we are? Many historians and philosophers today would agree that there is no such thing as the "past in-itself" or *"wie es eigentlich gewesen ist"* (Ranke),[3] that is, no talk about the past that is not already an interpretation. If we can reconstrue how the past is described, for instance, as Walter Benjamin does by writing history from the point of view of the victims rather than the victors, then we have also changed the past. But then, what could statements about the past be true of if the past is not real? The past must somehow *anchor* the present. One must accommodate realist intuitions about the past even if one's view of the past is always interpretive. There are phenomenological features of the past that get covered over by the antirealist stance. If one can change the past at will simply by reinterpreting it (a view that I refer to as *interpretive voluntarism*[4]), then it can no longer serve as an anchor for the interpretation of the present. We want our interpretations of the past to be taken as true, and not as the result of a voluntaristic rewriting.

The noted philosopher of science Ian Hacking prefers a more circumspect approach. He hesitates to say that the past can be changed. Instead, it seems safer to him to say that the *description* of the past can be changed. In *Rewriting the Soul*, Hacking writes,

As a cautious philosopher, I am inclined to say that many retroactive redescriptions are neither definitely correct nor definitely incorrect. . . . It is almost as if retroactive redescription changes the past. This is too paradoxical a turn of phrase, for sure. But if we describe past actions in ways in which they could not have been described at the time, we derive a curious result. For all intentional actions are actions under a description. If a description did not exist, or was not available, at an earlier time, then at that time one could not intentionally act under that description. Only later did it become true that, at that time, one performed an action under that description. At the very least, we rewrite the past, not because we find out more about it, but because we present actions under new descriptions.[5]

What makes this caution seem so sensible is that it avoids overstatement. It also recognizes that once choices are made and concrete actions are taken, there is no way to redo those choices and actions. They become an aspect of the world, which has a determinate history.

This way of addressing the issue does not foreclose debate, however. Those who are sympathetic to Hacking's project could take his arguments in different ways. For instance, a realist about the past might see Hacking as an antirealist who is attempting to undermine the distinction between the way the past is "in itself" and the way that it is "for us." By pointing out that there is no access to a thing except through the description of it, Hacking seems to the realist to be saying that if we change the description, we necessarily change the thing itself.

Hacking also says, however, that we cannot change the past simply by wanting it to be different. Initially, the argument assumes that there is no epistemic access to the thing independent of its description. Thus, it seems natural to infer that there is no ontological difference between the thing and its description. This is the quasi-nominalist interpretation of the initial assumption. The thoroughgoing nominalist will bite the bullet and say that if there is no determinate difference between the thing and its description, there is no such thing as the thing in itself independent of the description.

The thought that a change in the description entails a change in the thing itself depends for those with strong realist intuitions on positing what is denied, namely, the distinction between the thing and its description. That is, realist intuitions lead to the admission that if only the meaning changes but not the things or the events in themselves, then the distinction between the "in-itself" and the "for us" still obtains. Realists tend to argue that in order to be able to say that the description changes, there must be something that persists that is independent of the description. Otherwise, different descriptions would be of different things (events or states of affair), and there would be no way to say that we had a case of a different description of the same thing.

Despite Hacking's own critique of scientific realism in various forms, he clearly has strong intuitions that make him loath to adopt a nominalist or even a pragmatic stance. For Hacking this latter stance seems idealist in that it refuses to posit a reality of which the description would be true. The point is not simply that the description of the past changes, but that the world as it appears to different "styles of reasoning" will also change. Hacking does not want to lose the world to an argument that would say that there are as many worlds as there are interpreters. This is the position that he calls, in an essay entitled "Language, Truth, and Reason" in *Historical Ontology*, "subjectivism." There he distinguishes subjectivism, which he thinks is "inane," from "relativism" in a good sense.[6] Subjectivism maintains that "by thinking, we make something true or false." He counters it with what he calls "relativism" in a good sense, which asserts, "by thinking, new candidates for truth and falsehood may be brought into being." Hacking is thus theorizing the heuristic production of hypotheses rather than determining what makes a claim true or false.

Hacking identifies himself as an "anarcho-rationalist" who tolerates other views while maintaining the discipline of the standards of truth and reason inherent in his own style of reasoning. I take it that a style of reasoning is larger than a particular person's application of it, and is not simply put on or taken off in the manner of Plato's cloak. One finds oneself always already employing a style of reasoning. Even those authors whom Foucault identifies as "founders of discursivity" find themselves adhering to the rules and possibilities of the discourse they were the first to generate.[7] Nietzsche, Marx, and Freud represent not individual subjectivities, but generalizable styles of reasoning, even if (unlike in normal science) the founders retain an authoritative, and therefore problematic, relation to later developments in the field. A style of reasoning does not confront reality so much as it allows what counts as reality to appear. Such a style must be open to other possible views of reality than its own. As Hacking writes,

We cannot reason as to whether alternative systems of reasoning are better or worse than ours, because the propositions to which we reason get their sense only from the method of reasoning employed. The propositions have no existence independent of the ways of reasoning towards them.[8]

Perhaps one way to put the historico-ontological assumptions behind this analysis is to say that the reason there are different descriptions or interpretations of the world is not because there is no independent world, but because there is an independent world and it is infinitely complex. Human interpretations, therefore, will only ever be capturing a part of what could be the case. Furthermore, because the world includes the social, and the social involves change, the world is constantly changing. To describe the world will thus require different styles of reasoning as time goes by.

Another way to think about the issue of whether the past can be changed is to approach it through a distinction drawn by Heidegger. The past becomes what Heidegger calls our "facticity." There are facts about ourselves that it does not seem possible to change. Heidegger, however, distinguishes between facts that are true about people and facts that are true about things. He reserves the term "facticity" for the former and he uses the term "factuality" for the latter. Scientific realism would be a matter of factuality. A Heideggerian could well maintain that facts about natural objects, whether observable or unobservable, are what they are and are captured in scientific explanations. History, however, does not consist of facts in the sense of factuality but of facticity. Facticity, on Heidegger's account, involves an openness to *possibilities*. "Factical possibilities," even those in the past, can still be open, in contrast to factuality, which is fixed and determinate.[9] Human beings thus always have some open possibilities that, if redrawn, will affect how we understand other aspects of ourselves. This distinction between facticity and factuality and the resulting concept of "thingness" itself depends on how we understand our ways of talking about the world, and on a particular set of interactions. In that respect, Heidegger's view could reconcile the Sartrean voluntarism toward the past noted initially by Hayden White with Hacking's caution that what is being changed is only the description of the past. The Heideggerian account, to be discussed in more detail below, suggests that taking one fork in the road rather than another changes the nature of the journey. Thus, where we go influences our understanding of where we have been.

Memory and Memorialization

The past manifests in the present through *memory*, but what is memory? Insofar as memories can be true or mistaken, memory seems like a form of ontic, psychological interpretation. Memories can be disclosive, however, insofar as they can show us not only how the present relates to the past, but also how the past can continue to guide the present. This is the task of *memorialization*, or as I shall also call it, Remembrance. Memorialization is not just memory, because its goal is not simply recall of past facts. Memorialization is an attempt to hold up the past to the eyes of the present so that the present does not forget the sacrifices of those who existed prior to the present and yet who still seem alive and pertinent to the self-understanding of the present. In short, memorialization is when we remember to remember.

An initial definition of memory could well be, therefore, that memory is the "consciousness of the past." What is then the relation of present consciousness to the past memory? William James remarks, as we saw in the previous chapter, "the feeling of past time is a present feeling."[10] That is to say, when we remember, we do not perceive something in the past. On the contrary, the memory is itself in the present, and thus it would seem that there must be something in the present experience that marks the content as past.[11] An explanation is therefore required of how experience is both *datable* and *directional*. James distinguishes between the retained past of the specious present and the remembered past of memory, and he maintains that the experience of the specious present as fading into the past is not the same as memory proper. Retention is not memory. Memory brings back or "reproduces" an event that has completely faded out. When memory recollects a present that is now past, that past present will include its own sense of what was for it the immediate past. For James, insofar as the specious present is not simply the present moment of the Now, but also includes the fading of the past present and the dawning of the future present, it allows any creature, even one with no conscious memory, to distinguish now from then, or before from after. James's argument is roughly that if there is no temporality, there is no memory. Furthermore, if there is memory, therefore there is temporality. But his argument is not that if there is temporality, therefore there is memory.

To address these issues of datability and directionality, as well as of the relation of memory to the present, in more detail, let me now take up some lessons learned from the history of phenomenology, first German, then French. These vignettes will provide some conceptual distinctions that will help with the larger questions. I suggest in the second part of this chapter that Henri Bergson's philosophy provides an unusual account of these issues, one that deserves more attention than it has received of late.

Twentieth-Century German Phenomenology

Unlike James, Husserl thinks that retention is a form of memory, or at least he calls it "primary memory." Primary memory is different from what Husserl calls "secondary memory." Secondary memory, or memory proper, occurs when the event is entirely over and is brought back in recollection. Husserl senses that the term "primary memory" can be misleading. Indeed, even our contemporary distinction between short-term and long-term memory seems to be a distinction between two types of secondary memory, not between primary and secondary memory. Primary memory presents or perceives directly, whereas memory proper is precisely not a presentation. The retention is still part of the experience, even if it is not the "primal impression." Hence, Husserl speaks of memory's content as a *re*-presentation.

To make this point about the structure of primary memory or retention clearer, let me discuss an extended note from November 10–13, 1911. This date indicates that the note represents Husserl's more considered reflections about temporality, even if only part of the note was published.[12] One sign of progress is that retention is viewed more dynamically, with more attention paid to the "running-off" character of retention as it "sinks back" and fades from consciousness. As a temporal event recedes into the past, Husserl says that it "contracts" as the duration "flows off."[13] Take my example from chapter 2 of the opera singer who holds an amazingly high note for an astonishingly long time. At any point in my hearing of the tone, I am conscious not only of the tone, but also its duration. I am also aware of the difference between the flow and my perception of the flow. If, for instance, I nodded off at that point and awoke only when the singer stopped for a breath, I would know that the tone I heard before and after my moment of snooze was the same. I would not know exactly how long the note had been held, but I would be aware of the difference between objective time that went on uninterrupted and the subjective time with the slight snooze. To use the Husserlian terms explained in the previous chapter, my transverse intentionality would still be in place despite a slight gap in my longitudinal intentionality.

In a simpler version of his time diagram, he represents the flow as we see it between points A and E in figure 3.1. For Husserl the running-off phenomenon explains the continuity of the past: "I am

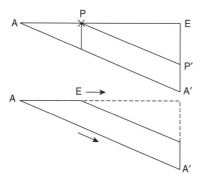

Figure 3.1
Husserl's time graph. Redrawn from Edmund Husserl, *On the Phenomenology of the Consciousness of Internal Time (1893–1917)*, trans. John Barnett Brough (Dordrecht: Kluwer, 1991), p. 29.

conscious of a continuity of time-points as 'immediately past' and of the whole extent of the temporal duration from the beginning-point up to the now-point as elapsed."[14] If points A and E represent "nows" that succeed each other, then this diagram brings out how they are distinguished from each other relationally. Because E contains retentions of retentions, its place in the series is distinct, and it could not have occurred anywhere else in the series. Husserl thus has accounted for how the continuity of the running-off modes of the duration is constituted. That is, he has accounted for the horizontal or longitudinal continuity of pasts on the basis of the transverse continuity in each moment, with its own array of retentions of retentions. The continuity of retentions then allows him to posit the unity of the flow. The unity of the flow also allows him to say that the flow can appear to itself: "the flow itself must necessarily be apprehensible in the flowing."[15] Here again we thus have an account of temporality temporalizing itself in a self-reflective way, prior to the emergence of the cogito. At least, that is what I take him to be positing when he speaks of "this prephenomenal, preimmanent temporality [that] becomes constituted intentionally as the form of the time-constituting consciousness and in it itself."[16]

This notion brings us back to the issue of the ontology of the past. There is a hermeneutical interpretation of memory and the past that Husserl's account opens up, even if he does not explicitly adopt it. His account maintains that each retention comes along and adds to the stream of consciousness. This point suggests that experience will change as each additional input conditions and changes the self-interpretation of what is already stored. Husserl thus recognizes that there is a "retroactive effect" whereby the new moment "modifies the reproductive possibilities for the old."[17]

If Husserl is right on this point, then there is no ontological issue about changing the past. Phenomenologically it is evident that with the addition of new experience, the past is always changing. Similarly, memory is in flux because conscious life is in flux.

Husserl writes, "Memory flows continuously, since the life of consciousness flows continuously and does not merely piece itself together link by link into a chain."[18] The point is that as each present is succeeded by another present and "sinks into the past," its retentions amount to a modification of other retentions from previous presents. This modification of the earlier retentions by later retentions opens up the possibility that what is considered as the "past" changes with the accumulation of successive retentions.

The historical Husserl would be wary of this pluralistic way of interpreting his diagram of time. Husserl thinks that objective time is constituted precisely by the accumulation of retentions. "Time is fixed," he notes, "and yet time flows. In the flow of time, in the continuous sinking down into the past, a nonflowing, absolutely fixed, identical, objective time becomes constituted."[19] Heidegger, in contrast, offers his own account of how objective time is constituted as a leveling off of primordial time, and he also accepts the Einsteinian idea that "time is always local time."[20] Let me provide some background for this point by looking more closely at Heidegger's account of the ontology of the past.

Heidegger challenges Husserl by moving from transcendental phenomenology to hermeneutical phenomenology. He takes over from Husserl the three-layered analysis of temporality into protention, retention, and primal impression, which Heidegger calls *ecstases*. If the guiding question here is whether the past can be changed, that question becomes more difficult in Heidegger, who distinguishes two senses of the past: *Vergangenheit* and *Gewesenheit*. If Heidegger did not completely reject the subject–object distinction, I would say that the *Vergangenheit* is the objective correlate to the subjective experience of *Gewesenheit*. The question then becomes twofold: can the *Vergangenheit* change, and can *Gewesenheit* change?

I will be interpreting these concepts freely, and I begin by noting that if this distinction occurred in English, one would be strongly compelled to say "the" *Vergangenheit*, but simply, *Gewesenheit* because the former term suggests a present-to-hand thing and the latter a range of experience. *Vergangenheit* is the past reified into present-to-hand objects and their relations to all the other objects. *Vergangenheit* is over and done with, and it sits at a distance from us because there is little we can do to change it.

Gewesenheit, in contrast, is the past as it is still effective in the present. *Vergangenheit* is expressed by tenses such as the simple past, whereas *Gewesenheit* is captured best in the imperfect tense. Insofar as *Gewesenheit* is still working itself out and its meanings are not yet established for certain, *Gewesenheit* is less a matter of reflective cognition, and more of a prereflective or skilled manner of coping directly with the world. *Gewesenheit* is also not over and done with, but involves the future.[21] We find ourselves always already in a situation, and thus our choices are not infinite, but limited. We do not simply adopt this situation, and we cannot reinterpret it voluntaristically. Instead, we are *thrown* into this situation and we have to "deal with it" as opposed to being "in denial" about it. If Heidegger had this language, he would more than likely say that "dealing with it" is authentic (if ontic), but that being "in denial" is inauthentic.[22]

Distinguishing these two senses of "past" puts Heidegger in a better position to deal with the question of whether the past can be changed. An objectivist will tend to construe *Vergangenheit* as the entire set of "all the facts." On such a view, the facts are fixed and the *Vergangenheit* cannot be changed.[23] For the objectivist, then, there is only one right account of the past. In contrast, for Heidegger the past is not frozen into facts. Furthermore, the very idea of "all the facts" is unintelligible. On the Heideggerian view, then, accounts of the past are always selective, and different selections of facts can tell different stories. Moreover, what counts as a fact is a matter of the style of interpretation. Different kinds of interpretation would thus result in different kinds of historical "reality."

As I mentioned previously, Heidegger recognizes two different senses of understanding, ontic and ontological. These correspond to the distinction in *Being and Time* between *Auslegung* and *Interpretierung*. Furthermore, they also draw on the distinction between two senses of truth, namely, ontological disclosure, or *Erschlossenheit*, and ontic discovery, or *Entdecktheit*. To capture this double sense in which the past anchors the present at the same time that the present can re-view the past (i.e., view the past again and form a different opinion of it), let me distinguish two different senses of "interpretation." On the one hand, there is interpretation that is done deliberately and reflectively. Interpretation in this sense of "*ontic interpretation*" can get things wrong and draws on its object domain for verification of its various claims. Deeper than this is interpretation in a different sense. At this level of "*ontological interpretation*" things either show up or they do not. Ontic interpretation can be right or wrong about the things it encounters, but to encounter things at all, there must first be a disclosure in an ontological interpretation of the world and the things that there are. Ontic interpretation picks out features of things within the world, whereas ontological interpretation discloses the world as such. Ontic interpretation discovers, and it can be true or false. Ontological interpretation discloses the context or the world, and this makes ontic interpretation possible.

I have been talking as if memory and memorialization were thematic, conscious, and reflective ways of interpreting the phenomena. That is *ontic* interpretation. But prior to ontic interpretation there is ontological interpretation, which allows the situation to be experienced or undergone in the first place. Although there is no experience that is not interpretive, ontological interpretation is built into the body in such a way that it conditions how we perceive our choices, our selves, and even our self-understandings. Although we can try to change our ontological interpretations, doing so is much more difficult than changing our ontic interpretations. Simply by moving our heads we can change what objects we see. But there are features of the way that we perceive that we do not alter in that action. The disclosure of the world is not itself an object within the world, and thus, bringing it up to a conscious level where it can be reflectively grasped will only ever be partially successful. Psychological, ontic interpretation cannot change the *Vergangenheit*. However, hermeneutical, ontological interpretation recognizes the normative and therefore the *political* dimension of our situated Being-in-the-world, our *Gewesenheit*.

Heidegger's student Hans-Georg Gadamer follows his teacher and insists on the connection with tradition as the basis for identity formation. On his view, whether we know it or not, our understanding is always influenced by the history of intervening interpretations. That is, when we read Kant, other philosophers (such as Hegel, Schopenhauer, or Nietzsche as well as a myriad of Kant scholars) who have developed their views as a reaction to Kant's philosophy influence our interpretation of Kant. Gadamer, the founder of contemporary hermeneutics, calls this history of influence *Wirkungsgeschichte*. This is a claim about the nature of understanding, and it implies that understanding is always interpretive. Gadamer's hermeneutical philosophy entails that there is no direct, unmediated read of, for example, the author's intention in the case of literary interpretation, or of the framers' intentions in the case of Constitutional interpretation.

In addition, however, Gadamer turns this ontological point into a normative prescription. He wants to account not only for understanding in general, but also, he wants to explain what makes some understandings better than others. "Better understanding" does not mean simply "more factual." For Gadamer, all understanding is also self-understanding. Therefore, an understanding that is aware of the intervening history of interpretations, and of the influence of these on the present interpretation, is better than one that ignores this history of influence. He thus calls for interpretation to reflect on itself and to become self-critically aware. His term for this self-critical awareness is *wirkungsgeschichtliches Bewusstsein*. His methodological analysis thus has the consequence that interpretation is never simply a description because it is always influenced by the history of intervening interpretations. Furthermore, insofar as every interpretation is at the same time a self-interpretation, it is better for interpretation to be more self-conscious of how it has been influenced by previous interpretations.

As a result of this emphasis on the importance of tradition, Gadamer is sometimes interpreted as being a conservative. His intention, however, is to say that in acting toward the future, we must have some sense of who we are. Tradition is what gives us that sense of ourselves. Gadamer is saying more concretely what Heidegger said above, namely, that an authentic relation to the future requires an authentic relation to the past. These relations are ongoing, with a movement in one direction requiring changes in the other direction as well. Gadamer's theory thus would indeed be conservative if one did not understand that what counts as the tradition can change. Gadamer thinks that we can change what we take the tradition to be by finding out that there are other possibilities in the past that were lost sight of, but that can be rejuvenated and restored. What counts as the tradition is always revisable. Tradition therefore is not necessarily reactionary, but it can be radicalized as well. Tradition is not synonymous with conformity. Historical memory can show us where the tradition was misinterpreted and where a people's self-understanding went awry.

Twentieth-Century French Philosophy

French poststructuralist philosophers are often construed by their critics as attacking continuity and coherence. The poststructuralists are thereby said to be incoherent defenders of irrationality. In their defense, I will point out that it is sometimes appropriate to call both continuity and coherence into question. For instance, "experience" is often theorized as if it were a unified whole with no gaps. To come to the issue of this chapter, when Husserl theorizes memory, he seems to build into his model the thought that memory is continuous, and that overall, experience flows along in a single stream. That is not the way experience feels from the inside, however. Especially if we think about the troubles we all experience with memory, we might well want to ask, what if discontinuous experiences were as typical as continuous experience? How reliable is memory? And is the unreliability of memory necessarily bad?[24]

These questions reflect the influence on French poststructuralism of Nietzsche, who points out that forgetting is a necessary feature of getting on with daily experience. If such forgetting did not take place, we would have too much to keep in mind all the time. Proust is also in play here, and Derrida reminds us that

memory, for Proust, far from being total and continuous, is intermittent and discontinuous. Our memories are out of our control. We remember only what our memories, acting on their own, happen to think it worthwhile to save.[25]

For Derrida "memory *is* or rather *must*, *should* be an ethical obligation: infinite and at every instant."[26] Memory could not be an ethical obligation if we had perfect recall. Thus, when memory is seen as a matter for practical philosophy rather than only for theoretical philosophy, it becomes equally important to take into account the gaps and disconnects in memory as well as its coherence and continuity. Even if what we forget is not in our conscious control, we still have obligations to remember certain things that tie us to others. What we forget can be as significant as what we remember. In the ethical sphere, then, the ideal of temporal continuity through retention becomes more problematic. Perhaps psychologists could gather some data on this question of the continuity of temporality. But to assume that all experience flows smoothly into the past and into memory is to forget the many lapses of memory that occur to most of us (eidetic memories apart), even if we are not yet aged enough to have "senior moments."

If memory has gaps, so will what we think of as the "tradition." The tradition is not everything that has happened in the past, but those features that have continuing influence and effects, often below the threshold of the visible. One theorist who has tried to make these subliminal features more visible is Pierre Bourdieu. I will turn to his account after first discussing the grandfather of recent French philosophy, Jean-Paul Sartre.

Jean-Paul Sartre (1905–1980)

Sartre's phenomenology of the past turns on what he considers to be a paradox: I cannot think of myself as being without a past, yet I am the one through whom the past comes into being. In *Being and Nothingness* (1943) this paradox is defined as the contrast between freedom and situation. I necessarily exist in a situation, on Sartre's theory, but at the same time I am radically free toward the meaning of this situation. For Sartre, human freedom is an all-or-nothing, total phenomenon. On the one hand, he insists that we are who we are because of our situation, and the situation is essentially our past. The past, he is fond of saying, is *irremediable*. On the other hand, he also thinks that "the past is what it is only in relation to the end chosen."[27] I choose my ends through my relation with the future, and the future involves changing the past. The past is "that which is to be changed," says Sartre.[28] The argument is that insofar as "nothing comes to me which *is not chosen*," it follows that "we can see too how the very *nature* of the past comes to the past from the original choice of a future."[29]

So the existential paradox is that I am who I am made to be by my past at the same time that I am the one who chooses the past. Sartre does not mean to say, however, that I can reflectively and consciously decide or deliberate about how to change my past. The past is changed only through action: "by action I *decide* its meaning."[30] What Sartre means by to "decide" here is not to "deliberate." For Sartre deliberation is always too little, too late. Deliberation gives the illusion that I am reflectively deciding, but in fact, the decision has already been made, and deliberation is only a way of temporizing or putting off the inevitable.

Sartre's theory thus combines the Bergsonian point (to be explained shortly) that the past does not cease to exist but that it becomes the context for present action, and Heidegger's point that we always project our possibilities toward the future. In Sartre's words, "it is the future which decides whether the past is living or dead."[31] For Sartre, therefore, there is a close connection between the historical and the temporal: "If human societies are historical," he writes, "this does not stem simply from the fact that they have a past but from the fact that they reassume the past by making it a *memorial*."[32] If temporality implies memory, history implies memorialization. How we see the past, whether as cohesive or as chaotic, will be decided by whether we see the past as continuous with the present or as "a discontinuous fragment."[33] Both of these are legitimate inferences about the past, but once in place, from that moment on the past "imposes itself on us and devours us."[34]

So the power that the past has over us is coming not from the past itself, but from the future. What does Sartre mean, then, when he says that my past "sinks into the *universal past* and thereby offers itself to the evaluation of others"?[35] To answer this question, we must also ask what is meant by the universal past. Furthermore, we should figure out what we mean when we talk about *the* past. Sartre offers in an earlier chapter of *Being and Nothingness* an explanation of how we get from the idea of my personal past to that of the universal past. Sartre explains that "universal temporality is *objective*."[36] The past of the world, or objective time, is deduced from the personal past, or temporality. Insofar as I view myself as an in-itself or a thing among others rather than as a for-itself or a conscious, free being in the world, I reduce qualitative temporality to quantitative, homogeneous instants strung together in a line. I have to give up my transcendence, my freedom, and affirm my facticity in order to see that "there is only *one* Past, which is the past of being or the *objective* past *in* which I was."[37]

Because of the power that the universal past has over me, insofar as it takes away my freedom and leaves me determined by the facticity of my situation, I flee it. Sartre sums up his account of the past by saying, "It is through the past that I belong to universal temporality; it is through the present and the future that I escape from it."[38]

From this quotation we can infer Sartre's abhorrence of time. Time is our situation and our facticity. Insofar as Sartre posits us as being radically free, he will always be resentful of our temporal condition. At the same time, he sees the necessity of being in time, and he formulates this necessity in a temporal version of a Kantian moral imperative. Expressed hypothetically, the imperative would be "If you wish to have such and such a past, act in such and such a way."[39] Action is the only way to redeem ourselves and to prove our freedom. On Sartre's analysis, however, we are our past, and the past is irremediable. We are condemned to having to try continually to do the impossible. *Being and Nothingness* thus concludes with his pessimistic view that humanity is a futile passion.[40]

Pierre Bourdieu (1930–2002)

For the French sociologist and philosopher, Pierre Bourdieu, Sartre's emphasis on the body is a step in the right direction. Sartre's theory of radical freedom is, however, a mistake. Bourdieu sees us as conditioned by our bodily habits, the many ways our body gets molded by the subliminal socialization practices that we experience in our upbringing. His views about temporality are thus closer to Heidegger's than to Sartre's, except for the major caveat that Heidegger's privileging of the future is a sign not of authenticity, but of inauthenticity. Bourdieu thinks that Heidegger is misrecognizing the source of the problem, which is not the future but the past. Where Heidegger sees possibilities coming into the present from *the future*, Bourdieu sees the reproduction in the present of *past* objective structures. These structures are built into the body, or more precisely, into the set of bodily dispositions that Bourdieu calls the habitus. The habitus is situated in a social field and participates in it in much the same manner as one participates in a game. To give one of Bourdieu's favorite examples, in games such as football or soccer or basketball, the player sends the ball not to where the other player is at the moment, but to where the other player soon will be. A successful anticipation, however, depends on long hours of practice that enable the teammates to anticipate the bodily capabilities of one another. What seems like a successful instance of looking ahead is really, then, a case of implicit looking-back. In *The Logic of Practice*, Bourdieu inverts Heideggerian temporality by seeing "the presence of the past in this kind of false anticipation of the future performed by the habitus."[41]

Although Bourdieu differs with Heidegger's account of the direction of time, I see Bourdieu as providing a concrete (although ontic) way of understanding Heidegger's gnomic injunction that temporality temporalizes. Bourdieu argues that time is not to be thought of as a thing or an object that a subject "has." Bourdieu wants to describe the actual practices of "agents," not "subjects." He therefore says that he is reconstructing the point of view of the acting agent, which is the point of view of "practice as 'temporalization,' thereby revealing that practice is not *in* time but *makes* time (human time, as opposed to biological or astronomical time)."[42] Bourdieu distinguishes further between the conscious project of the future (*avenir*), and the prereflexive protention of the "forthcoming" (*à venir*). Bourdieu is drawing here on Husserl's notion of the protention of, for instance, the hidden faces of a cube. One sees a cube, but one does not see all the faces of the cube. Nevertheless, one could not be seeing a cube as a cube unless one sensed the hidden faces, which are said to be "quasi-present." Similarly, unlike the future in the sense of *l'avenir*, which is a conscious projection of a future present, Bourdieu thinks that in the *à venir* time is not noticed even though it is there.

On my view, his account of the soccer players is not quite the correct phenomenology. He leaves out the fact that the players certainly do know the time. The players always have a sense (often more implicit than explicit) that the "clock is running." Bourdieu is right, however, when he says that time is first noticed consciously only when there is a breakdown: "Time (or at least what we call time) is really experienced only when the quasi-automatic coincidence between expectations and chances, *illusio* and *lusiones*, expectations and the world which is there to fulfill them, is broken."[43] How does a temporal breakdown occur? For Bourdieu, a breakdown in the *à venir* can be the result of a disruption of either the *illusio*, the expectations that are built into the habitus, or the *lusiones*, the probabilities that are built into the social field.

An example of such a breakdown is unemployment. Unemployment suddenly leaves one with time on one's hands. In contrast to the busy person who never has enough time, the unemployed become conscious of dead time, where time hardly seems to move at all. (Benjamin would call this "empty time.") Bourdieu thus adds a social explanation of Heidegger's distinction between the inauthentic person who never has enough time and the authentic person who always has time. Bourdieu sees this surplus of time as a sign not necessarily of authenticity, but more often of powerlessness. Consider the case of waiting. The powerful, whose time is valuable, make the powerless wait. On Bourdieu's analysis, waiting is a form of submission.

Even if Heidegger is right in suggesting that to have control of one's life is to have control of one's time, such that one can "take one's time," Bourdieu believes that there is more to say about how personal temporality derives from social temporalization. For example, the academic and the artist live privileged lives where the line between leisure and work is blurred and time seems to be bracketed. The quasi-free time of the academic resembles the negated time of permanently unemployed subproletarians in that both have time on their hands; but the subproletarian pays the price for the academic. For Bourdieu, the fact that both cases would satisfy Heidegger's criterion for authentic temporality amounts to a refutation of Heidegger's account by a putative reductio ad absurdum. More generally, this social contradiction allegedly casts doubt on any analysis of individual authenticity that ignores the social conditions of temporality that are built into the body. Bourdieu therefore concludes that "time is indeed, as Kant maintained, the product of an act of construction, but it is the work not of the thinking consciousness but of the dispositions and practices."[44]

Henri Bergson (1859–1941)

Gilles Deleuze is largely responsible for two waves of renewed interest in Henri Bergson that have occurred since the second World War. His book, *Bergsonism*, was published in French in 1966 and led to a short-lived revival of Bergson in France. He called the book *Bergsonism* rather than simply *Bergson* because Bergsonism was indeed a philosophical movement in its own right—just as much, anyway, as the movements of positivism and pragmatism that it opposed. In fact, Bergsonism was in its own time a much larger and more popular movement, inspiring major aesthetic and political changes as well. The publication of the English translation of Deleuze's book in 1988 led to a renewal of Anglophone philosophical interest in Bergson in the early 1990s. Bergson never regained, however, the international fame that he had in his lifetime (1859–1941). Since then his role in philosophy has been that of the crown prince who never became king. Nevertheless, his ideas about temporality when combined with Deleuze's appropriation of them are, in my opinion, some of the most unusual in the history of modern philosophy. A history of temporality that did not discuss Bergson's idea of *la durée* (duration) would be seriously remiss. In particular, discussion of Bergson's views about memory and the relation of the past to the present is indispensable in a chapter on the past.

Bergson via Merleau-Ponty

Before discussing Deleuze's revitalization of Bergson, however, I want to note that there was at least one other major twentieth-century philosopher who acknowledged Bergson's influence. In his inaugural lecture to the Collège de France on 15 January 1953, Maurice Merleau-Ponty spent a considerable amount of time discussing what he also called Bergsonism. Translated as *In Praise of Philosophy* (1963), Merleau-Ponty's intention in this lecture is not only to bow politely in the direction of his most famous predecessor in this chair of philosophy, but also to show the connection of Bergsonism to the type of phenomenology that was dominant in that period both in France and in the Anglophone world. At the same time, Merleau-Ponty also has to mark the differences between his own program and Bergson's. Even if Bergson was not a true (Husserlian) phenomenologist, Bergson's approach was sometimes identified as bearing a family resemblance to phenomenology. Merleau-Ponty also discusses Bergson at length in his lectures on nature in 1956–57.[45] In addition, there are several long footnotes on Bergson in the *Phenomenology of Perception* (1945), a short presentation from 1959 in *Signs*, some remarks in *The Visible and the Invisible*, and a close reading of *Matter and Memory* in lectures published as *The Incarnate Subject*. Reading Bergson through the parallax view obtained from the differing perspectives of Merleau-Ponty and Deleuze will result in a multidimensional view of his philosophy and its significance today.

In the inaugural lecture Merleau-Ponty glosses Bergson's view of consciousness in a way that resembles without being identical to Sartre's well-known view of the relation of consciousness and nothingness. Just as for Sartre, whom Merleau-Ponty often criticized, for Bergson emptiness or nothingness is introduced by the *pour-soi* or being-for-itself (consciousness) into being, or the full presence of the *en-soi* (the mute being-in-itself of external objects). Bergson shows, says Merleau-Ponty, that a nothingness in consciousness would be the consciousness of a nothingness, and that it would, therefore, not be nothing. But this is to say in other words that the being of consciousness is made of a substance so subtle that it is not less consciousness in the consciousness of an emptiness than in that of a thing.[46]

This quotation makes the point that nothingness comes from consciousness and that consciousness desires to be more than nothing, that is, to be what Sartre calls being *en-soi-pour-soi*. Merleau-Ponty goes on to read Bergson as a proto-existentialist who anticipates the Sartrean and Heideggerian analyses of Angst and vertigo: "if true philosophy dispels the vertigo and anxiety that come from the idea of nothingness, it is because it interiorizes them, because it incorporates them into being and conserves them in the vibration of the being which is becoming."[47] Here Merleau-Ponty is expressing Bergson's ideas in a way that shows Bergson to be of contemporary interest to Merleau-Ponty's audience, even if not of the same level of interest as Merleau-Ponty himself.

In the lectures on the philosophy of nature at the Collège de France in 1956–57, Merleau-Ponty clarifies further the relation of Bergson and Sartre. He confirms that they have similar views about the relation of being and nothingness:

Sartre had the idea that in the history of consciousness, there is not a preliminary lack: the human creates both the lack and his solution. Likewise, Bergson thinks in *Creative Evolution* that philosophers create problems and their solutions at the same time.[48]

Sartre and Bergson are said to occupy similar places in the history of consciousness because their conceptions of being and nothingness do not connect enough to allow for a conception of either nature or history. Merleau-Ponty sees Sartre's opposition between being-for-itself (subjectivity) and being-in-itself (objectivity) as precluding any account of the self-development of materiality. In other words, Sartre's early philosophy allows us too much freedom in the constitution of both our nature and our history.

For Merleau-Ponty the popular movement known in France in the early twentieth century as Bergsonism depends on a historically limited reading of Bergson, one that turns Bergson's fecund explorations into dogmatic doctrines. Merleau-Ponty thereby establishes the view that the early work of Bergson was his best because it was adventurous, and that the older Bergson became more solid but stolid. The early Bergson is standardly construed as the philosopher who discovered duration and intuition in contrast to modern science, which supposedly cannot grasp these important phenomena in its efforts to quantify and objectify everything. Duration is the idea that time is stretched out and not a series of atomistic nows. Duration is a temporal unity that can be longer or shorter. (Later French historians spoke of the *longue durée*, that is, of a history covering longer periods of time than could be perceived by individuals or generations.) Bergson was particularly interested in *la durée* because he thought that it was something science could not capture in its explanatory web.

For that same reason, intuition is also standardly viewed as a central phenomenon in Bergson's thought. Intuition is a primordial contact with things that is presupposed by and prior to language and conceptualization. According to Merleau-Ponty, Bergsonism is usually construed as emphasizing the idea of a subject that, before language, is "already installed in being."[49] Intuition is the true thought that is fused with things in a naive contact that genuine philosophy must rediscover. Intuition is said to be a "simple act," "viewing without a point of view," direct access to the interior of things unmediated by signs or symbols. Bergsonism is intensely antiscientific, and it reduces the younger Bergson's philosophy to some polemical theses articulated in the later stages of his philosophical development. Merleau-Ponty lists three such Bergsonian doctrines: (1) intuition is prior to intellect and logic; (2) spirit is restored to its primacy over matter; and (3) life is discovered to be more primordial than mechanism.

For Merleau-Ponty these polemical doctrines tend to overemphasize Bergson's break with received doctrine at the cost of missing what the younger Bergson really had to say. The mistake of popularized Bergsonism is to reduce Bergson's critique of the positivism of his time to some doctrinal principles that are frozen into concepts, not unlike those of the positivistic philosophers whom he was attacking. For Merleau-Ponty, the real Bergson did not see the true philosopher as lost in or absorbed by being. Rather, philosophy must experience itself as transcended by being: "It is not necessary for philosophy to go outside itself in order to reach the things themselves; it is solicited or haunted by them from within."[50] What "haunts" us most particularly is *la durée*. Bergson's claim is that the human ego is essentially duration and cannot grasp another being except as having its own duration.[51] One's own manner of using up time, for instance, is a choice of one possible *durée* drawn from an infinity of possible *durées*. Even the idea of the *durée* of the universe is only an extension of the entire length of one's own *durée*. The temporal experience of waiting for sugar to dissolve in an espresso, to use a typically Bergsonian example, requires one to posit the objective time that is required for the sugar cube to melt in it. There is thus for Merleau-Ponty's Bergson a complicity between the temporality of experience and the time of being.

In the lecture course of 1956–57 Merleau-Ponty suggests more critically that despite this complicity, Bergson cannot really explain the oneness of time. These criticisms are consistent with those mentioned in footnotes in the *Phenomenology of Perception*. There he says that Bergson "never reaches the unique movement whereby the three dimensions of time are constituted."[52] Merleau-Ponty identifies three errors in Bergson: (1) the body is only ever the objectified body, not the lived body; (2) consciousness is only ever explicit knowledge-about, and not the prereflective skills and awareness of the world; and most significantly for present purposes, (3) "time remains a successive 'now,' whether it 'snowballs upon itself" or is spread in spatialized time."[53] Alleging that Bergson cannot explain why duration is "squeezed" into the present, Merleau-Ponty repeats this criticism of the metaphor of time as a snowball in later footnotes as well. In one footnote he says that for Bergson it is *duration* that "snowballs upon itself," and thus Bergson is wrong to try to theorize time out of a reified, "preserved present."[54] In another footnote he says that it is *consciousness* that "snowballs upon itself" and tries to get everything into the present.[55] Bergson's "snowball" account of time is right, according to Merleau-Ponty in the *Phenomenology of Perception*, to emphasize the continuity of time over time. The continuity of time, he says, is an "essential phenomenon."[56] Nevertheless, Bergson's explanation goes awry because Bergson puts the cart before the horse and tries to explain the unity of time through its continuity. In my terms, the issue about the "oneness" of time at any given time is different from the issue about the "unity" of time over time. The issue about oneness involves the synchronic issue why any given experience is a single experience, where all the data fit together in a whole. The problem of unity involves the diachronic issue of how the various moments form a single consistent and coherent sequence of experiences. Merleau-Ponty thinks that Bergson has confused these two issues and thus has misdescribed the phenomenology of the past, present, and future. In particular, Bergson's snowball account, in its preoccupation with forcing everything to accumulate in the present, does not sufficiently explain how we distinguish whether a given moment is a past, present, or future moment.

In the 1956–57 lectures on nature, Merleau-Ponty raises the same issue in terms of Bergson's discussion of Einstein's relativity theory. In *Durée et simultanéité* Bergson notes, "we bring with the self, everywhere we go, a time that chases away the others, like the clearing attached to the walker, [and that] makes the fog back away with each step."[57] The example of waiting for the sugar cube to dissolve shows that duration cannot be completely interior. My inner sense of duration has an objective correlate, and I am forced to recognize a time that is other than my inner time. There are always at least two times. The question is, if for the experiencer time is necessarily one, how can the oneness of these two times be explained?

This issue is focused in Bergson's critique of Einstein, or more accurately, in Bergson's attempt to explain relativity philosophically. On Merleau-Ponty's reading of Bergson, the physicist must have a sense of experienced time in addition to the objective time that concerns physics. In order to be able to say that time is relative to the observers, the physicist must be "both faithful and unfaithful to his principle: faithful, since he links time to the instruments of measurement, but unfaithful, since he confuses the effectively lived time of the observer situated in S and the attributed time of the observer situated in S-prime."[58] I read Merleau-Ponty as saying that for Bergson the Einsteinian physicist must operate with at least two perspectives on time. The physicist then shifts back and forth between these perspectives without being aware of it. The Einsteinian physicist is thus a "two-timer," both because there are always at least two temporal standpoints involved, and because it seems necessary to deploy both of these contradictory standpoints at once, which is supposedly impossible.

According to Merleau-Ponty, Bergson's solution to the question of how many "times" there are requires understanding the differences between a physicist's and a philosopher's type of explanation. Bergson believes that the philosopher can think about time more successfully because the physicist "multiplies the successive egocentric views rather than bringing about the philosophical coexistence of the times of the different observers."[59] Bergson ties this distinction to the difference between an egocentric and an intersubjective account of the plurality of time. The physicist is trying to theorize the world from all points of view at once, but at best is doing so only successively, taking one point of view after another. Supposedly only the philosopher can see time as one. The philosopher can posit the world as a whole, that is, "a philosophical non-physical simultaneity."[60] Instead of the oneness of time being the outcome of a process of occupying one perspective after another and storing them in a single ego, the oneness of time is the result of a posited intersubjective simultaneity, that is, the sum of simultaneous perspectives on every facet of the world. Empirically the latter is a physical impossibility, but it is not a logical impossibility.

For Merleau-Ponty this solution does not really work. He believes that although Bergson is right to emphasize intersubjectivity over a methodological egocentrism, more is needed to explain why the sum of intersubjective perspectives should form a single whole. Intersubjectivity is required because there must be at least two perspectives for an objective simultaneity to obtain. But more than that is necessary, and Merleau-Ponty sees the fuller answer in the structure of perception. A minimal model of perception requires two conspecifics triangulating on the same aspects of the world: "It is because two consciousnesses have in common the extreme portion of the field of their exterior experience that their time is one."[61] In other words, no two people have the same view on things and "co-perception is not identical perception."[62] Nevertheless, there must be sufficient overlap between the two perceptual fields to be able to say that they are perceptions of the same world and not completely different perceptions. The argument is thus that without a parallax view, there could not be a common object. This argument about space is then applied to time: I have to be able to triangulate the difference between my temporal experience, the temporal experience of others, and the time of the public world to be able to know that these different perspectives are finally possible because the time in which we live is one, even if we all experience it differently.

We should perhaps not be surprised that the author of the *Phenomenology of Perception* thinks that Bergson's most fundamental philosophical insight is that the complicity of experience and being is grounded in perception. Merleau-Ponty sees his own emphasis on the present reflected in Bergson's *Matter and Memory*, where *durée* is defined as the emptiness of the past and the future in relation to the perceptual fullness of the present.[63] Merleau-Ponty believes that Bergson may not have grasped the full meaning of his own terms when he wrote, "Whatever the intimate essence of that which is and of that which happens may be, we are of it."[64] These words suggest first of all the evolution of humans from animality, the animal from cosmological consciousness, and cosmological consciousness from God. These ideas may date Bergson, and because of them his philosophy will remain that of an earlier time, such that efforts of revival will invariably seem to be living in the past. Merleau-Ponty reads Bergson, however, as meaning by the phrase, "We are of it," that what we perceive in all beings are the notions of matter, life, and God that are symbolic of our lives. Merleau-Ponty sums up Bergson's views on perception as follows:

The genesis which the works of Bergson trace is a history of ourselves which we tell to ourselves; it is a natural myth by which we express our ability to get along with all the forms of being. We are not this pebble, but when we look at it, it awakens resonances in our perceptive apparatus; our perception appears to come from it. That is to say, our perception of the pebble is a kind of promotion to (conscious) existence for itself; it is our recovery of this mute thing which, from the time it enters our life, begins to unfold its implicit being, which is revealed to itself through us. What we believed to be coincidence is coexistence.[65]

Merleau-Ponty thus gives us Bergson as a phenomenologist of the prereflective access to the world through what Bergson called intuition. Philosophy in Bergson's hands comes to much the same thing as it does in a phenomenologist's hands. The unity of experience and its temporal oneness are guaranteed by our rootedness in being and our (perhaps mythical) sense of being at home in the world. This is one reading of Bergson. Gilles Deleuze provides a contrasting one.

Bergson via Deleuze

In contrast to Merleau-Ponty's Bergson, who is an eminently cogent proto-phenomenologist, Deleuze's Bergson is an entirely different and more difficult proto-poststructuralist. In Deleuze's hands, the early writings of Bergson in particular anticipate what became the poststructuralist emphasis on difference rather than identity, multiplicity rather than unity, the virtual rather than the real, and pluralism rather than monism. Nevertheless, Deleuze does not sell Bergson's metaphysical side short, and he recognizes the attraction of the absolute oneness of temporality in Bergson's thought. The contrast between duration and the absolute is, finally, so basic that it becomes the crux of current attempts to resuscitate Bergsonism.

For present purposes let me concentrate this discussion on the problem of the past and how the past connects to the present. I will focus on Deleuze's book, *Bergsonism*, using his *Difference and Repetition* as a resource for understanding his own views as well as Bergson's. Deleuze's *Bergsonism* raises challenging objections to common assumptions about the relation of the past to the present. The problem is that these objections sometimes sound strongly counterintuitive. Consider the case of where memories are stored. For Deleuze-cum-Bergson, the question of *where* recollections reside does not arise. Or if it does, the answer is not that they are stored in the brain. The brain on Bergson's theory is just an image of a particular form of matter, and recollection-subjectivity cannot be reduced to matter. In *Matière et mémoire* Bergson refuses the claim that perceptions are stored in the brain; instead, he says, the brain is itself in the perceptions ("elles [les perceptions passées et présentes] ne sont pas en lui [le cerveau]; c'est lui qui est en elles").[66] "Recollection," says Deleuze, "therefore is preserved in itself,"[67] and Bergson himself says that recollection "preserves itself."[68]

The "therefore" in Deleuze's sentence, however, could be unpacked. Much like Heidegger's claim that temporality temporalizes itself, the claim that recollection preserves itself need not be taken as totally mysterious and paradoxical. We could say, for instance, that memory remembers itself. Certainly I can remember having remembered something on an earlier occasion. Although "remembering remembering" is harder than remembering the initial event, it is neither impossible nor unusual. I can remember having taken examinations where I did remember what I had learned earlier (as well as ones where I did *not* remember a particular lecture).

Remembering having remembered is not the same, however, as the memory of the act of memory. When I remember having remembered, I remember the circumstances and perhaps the content, but not the noetic mental act itself. There is no such psychological state, I contend, no second- or third-order memory of a first-order memory. Bergson is right, in my opinion, to say that remembering is much like understanding a sentence. When I hear the sentence, I know right away what is being said. Even if the sentence is not clearly audible or well articulated, I can still understand it. There is nothing mysterious about my capacity to understand the particular sentence. I do not have to invoke explicitly the rules of grammar or the laws of phonetics to understand the mumbled sentence. The grammar is built into the process of understanding. Furthermore, the grammar is not psychological. That is, it does not appear to me psychologically. Rather, just as I see the lemon on the tree and not in my mind, I understand the sentence without any psychological awareness of how that understanding works.

Before going much further, it is crucial to understand that for Deleuze, to be is not to be present. Much like Heidegger, he thinks that a good part of the problem of understanding the phenomenon of memory, and thus, the relation of past to the present, stems from confusing Being and being-present. Instead of saying that the present is, Deleuze thinks that it is better to say, "the present *is not*; rather, it is pure becoming, always outside itself. It *is* not, but it acts."[69] To reduce everything to being-present leads to two common assumptions about the past that Bergsonians challenge: "On the one hand, we believe that the past as such is only constituted *after* having been present; on the other hand, that it is in some way reconstituted by the new present whose past it now is."[70] Deleuze characterizes these assumptions as "psychological." They misconstrue the temporality of the past because they think of time only from the standpoint of succession. On this psychological model, each present enters into the past as it is superseded by a new present. Bergsonians like Deleuze will ask, however, if to be real is only ever to be present, then where does the old present go? Deleuze quotes Bergson from *L'Energie spirituelle* (1919), where Bergson explains the relation of present perception to recollection or memory as follows:

> I hold that the *formation of recollection is never posterior to the formation of perception; it is contemporaneous with it.* . . . For suppose recollection is not created at the same moment as perception: At what moment will it begin to exist? . . . The more we reflect, the more impossible it is to imagine any way in which the recollection can arise if it is not created step by step with the perception itself. . . .[71]

The claim about the relation of the past to the present that is uniquely Bergsonian is, then, that the past is not ontologically distinct from the present but is simultaneous with it insofar as the two temporal dimensions "coexist." Bergson has employed a standard aporia about successive time to argue for a conception of temporality as simultaneous. Every present is already imbued with its past, and the past is really part of the present. If that is right, then we are always living in the past, if only because the past is contemporaneous with the present. The past "coexists" with the present. "À vrai dire, toute perception est déjà mémoire," says Bergson, and then he adds with emphasis, *"Nous ne percevons, pratiquement, que le passé*, le présent pur étant l'insaisissable progrès du passé rongeant l'avenir." ("Every perception is already memory. *Practically, we perceive only the past*, the pure present being the imperceptible progress of the past nibbling away the future.")[72]

In *Difference and Repetition* Deleuze is faithful to Bergson when Deleuze first describes this "paradox of coexistence," which he states as follows: "If each past is contemporaneous with the present that it was, then *all* of the past coexists with the new present in relation to which it is now past."[73] Though not exactly a paradox, the claim that all the past is present if any past is included at all is indeed a difficult thought to digest. Deleuze explains the alleged paradox with an even more unsettling thought: "The past does not cause one present to pass without calling forth another, but itself neither passes nor comes forth. For this reason the past, far from being a dimension of time, is the synthesis of all time of which the present and the future are only dimensions."[74] The present is thus the past in its most condensed or "contracted" degree. This point amounts to an a priori claim about all time. Moreover, insofar as the entire past can never be accessed at once, it becomes necessary to speak of "the Past which was never present": "In effect, when we say that [the past] is contemporaneous with the present that it *was*, we necessarily speak of a past which never *was* present, since it was not formed 'after.'"[75] Was the past there "before"? Deleuze says that the past's contemporaneity with its present was "already-there, presupposed by the passing present and causing it to pass."[76] "The Past which was never present" is then said to play the role of ground in Bergson's account of temporality.[77]

Bergson's way of illustrating this coexistence is to draw a cone. The tip of the cone is the present. The present is said to be temporality in its most "contracted" form. As the point at the tip of the cone is expanded by decompressing or relaxing it (*détente*), the present and past relations become more apparent and can be noticed. This notice can take the form of recollection, if the relations are remembered, or perception, if things are taken as being directly given. As I understand it, the argument is that recollection must always be possible, or otherwise we would not know that

perception was perception. The difference between them is required
for perception to be possible.

Given Bergson's distinctive way of thinking about the relation
of the past and the present, how does his model of the cone compare
to Husserl's linear diagram? Bergson and Husserl at first glance
appear to be offering two conflicting models of time. (See figures
3.1 and 3.2.) Metaphysically, they would even seem to be contradicting each other. Husserl sees time as successive, and has a
problem explaining simultaneity, namely, how the past can hook
up with the present. Bergson sees time as simultaneous, and has a
problem explaining succession, namely, how to tell the difference
between the past and the present if the past is simultaneous with
the present.

Taken as metaphysical theses, then either time is successive or
it is simultaneous. If it seems that both of these theses could not
be true at the same time, nevertheless each would seem to have a
legitimate claim on the theorist and both must be held. Thus, in

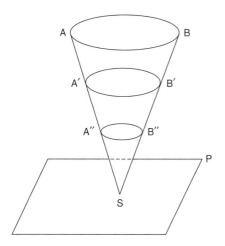

Figure 3.2
Bergson's time cone. Redrawn from Henri Bergson, *Matière et Mémoire* (Presses
Universitaires de France, 1939), p. 181.

faithfully espousing one of these theses, each philosopher must at the same time consort with the other. The apparent paradox then leads to further problems, such as explaining how there can be no change without time, but how time itself does not change.

The best way to make sense of Husserl's and Bergson's theories of time, as I read them, is to understand them as phenomenologies of temporality, not as metaphysical explanations of universal time. Phenomenologically, Husserl is explaining time-consciousness diachronically, and Bergson is explaining it synchronically. There is no contradiction if what is at stake is the phenomenology of temporality, because both explanations are sound. Husserl explains temporal flow as it goes by, and Bergson explains how the past is present in present experience precisely as past. That is, Husserl's diagram suggests that the relation is between two moments of time, where one is retained in the other. Bergson's cone, in contrast, captures how the past is meaningful only from within the present.

If this way of reading Bergson makes sense, then let me go on and consider Deleuze's reconstruction of Bergson's account of memory, which Deleuze claims is often misunderstood. To understand memory as Bergson does, it is first necessary to think through his claim that the past "coexists" with the present. To agree with this notion would require us to abandon some common understandings of the past. If we accept the Bergsonian idea that duration involves the past and the present not as two successive moments, but as two elements that coexist, then we must give up at least four commonsense assumptions about memory and the essence of time. We have to give up the "badly analyzed composite" whereby we assume wrongly that:

(1) we can reconstitute the past with the present; (2) we pass gradually from one to the other; (3) that they are distinguished by a before and an after; (4) that the work of the mind is carried out by the addition of elements (rather than by changes of level, genuine jumps, the reworking of systems).[78]

Deleuze thinks that the mind works more by leaps and abrupt changes than by smooth reasoning or gradual transitions. Discoveries and insights are the result not of seamless thinking-through, but of sudden concatenations and shifts. Bergson speaks of memory as a leap into the past. The question is, which past is it that one leaps into? Deleuze distinguishes the "pure past" or the "past in general" from any and every particular past. The past in general is the capacity to understand or experience a particular past. It too is not psychological. Bergson speaks of the "leap into ontology," and Deleuze says,

It is a case of leaving psychology altogether. It is a case of an immemorial or ontological Memory. It is only then, once the leap has been made, that recollection will take on a psychological existence.[79]

At this particular point in the text, however, Deleuze is trying to do justice to Bergson's cone, with its explanation of the relation of past and present as being not two separate moments, but as all our past coexisting with each present. If access to the past is to occur in the present, then there is no issue about the reality of the past and nothing mysterious about recollecting it. The Bergsonian thesis is that "the past does not follow the present, but on the contrary, is presupposed by it as the pure condition without which it would not pass."[80] That is, if there were no past, there could be no present. Of course, it could be equally true that if there were no present, there could be no past. Of these two inferences, whereas probably the majority of philosophers would accept the second, an equal or greater majority might find the first either unintelligible or false. The first, eminently Bergsonian inference is thus philosophically more interesting.

What, then, are some issues about this peculiar Bergsonian analysis of the relation of past and present? One might think that there is some circularity in the claim that the past is presupposed by the present, given that the past is said to be part of the present. For if the present contains the past, then all that is being said is that the present presupposes the present.

There are two replies that could be made to this objection. The first is to point out that Deleuze says that it is the past as "pure condition" that is presupposed. I will come back to this point immediately after presenting the second reply, which is the following. To say that the past "co-exists" with the present is not to say that the past is coextensive with the present. There must be a difference, but the difference is less problematic if the past and present are not separated by either temporal lag or logical aporia. Thus, when Deleuze says that the past and present are not distinguished by a before and after, I take it that he means not only a temporal before and after, but a logical one as well. If the past and present coexist, then no experience could be introspected directly to tell us whether it was a case of perception or recollection. As was the case for Kant, the before and after would not be built into each particular experience, but would be determined by the ordering of the mind. Deleuze suggests that a "recollection of the present" is possible, but that it would be a dysfunctional *paramnesia*.[81] Deleuze emphasizes instead the idea of a "past in general," or a "pure past," which is not a particular past but a condition for the possibility of the "leap into ontology." The idea is that the past is fixed and cannot be changed. The past is our "facticity," and goes with us as we move on. The past accumulates new features as we have new experiences and make certain decisions, but it is always with us as the context within which our self-interpretations arise and from which they have any intelligibility at all.

Deleuze contra Bergson?

In this section I return to the issue raised by relativity theory, namely, if time is only ever local time, how many times are there? Bergson is, at the end of the day, a monist about time. Bergson believes that there is only one time, and one time only. Deleuze, in contrast, as a poststructuralist, might well be expected to be a temporal pluralist. Without going too far into Deleuze's later views, we can see him wrestling with this tension in his treatise on Bergson. He may well be wrestling not so much with Bergson as with himself, or perhaps, his inner Bergson.

Insofar as the topic of this book is not time, however, but temporality, there may not be a need for a winner in this wrestling match. A possible solution is to say that from the standpoint of temporality we never directly experience universal time, but only our own sense of temporality, and thus, only local time. But it could equally well be said that we never experience time as merely local. When the space cadets are cruising the cosmos on their space yacht at nearly the speed of light, they do not experience time any differently than the people remaining on Earth do. When the space yacht gets back to Earth, the youthfulness of the cadets can be seen, but the temporal experience was, I speculate, not perceived any differently in either case.

Asking the question in the following way might best make Bergson's case. When the space yacht gets back from its trip around the cosmos at nearly the speed of light, presumably the space cadets' watches will read differently from those that have stayed on Earth. Could one then ask, what time is it really? The Bergsonian intuition is that there must be a single time that is the right time. In contrast, relativity theory could answer that there is only ever local time. This implies that the space cadets will probably have to reset their watches to Earth time, but that decision is only a matter of convention.

Again, however, that is an issue about public, objective time, which is equally a convention. A better response would thus be to say that the Kantian intuition is correct that I always experience time as "One." At the same time, Einstein could be right that there is more than one "One." The phenomenologist of temporality does not have to settle this dispute, Bergson could have said, because it does not show up in experience. What Bergson did say, however, is that time *is* one, despite differences in duration. How is Deleuze going to present that metaphysical view persuasively, given poststructuralism's allergy to metaphysics?

In *Difference and Repetition* Deleuze stages this issue as a confrontation between Nietzsche's eternal return and Bergson's insistence on the unity of time. Or at least I infer that he could have Bergson in mind when he writes,

It is said that the One subjugated the multiple once and for all. But is this not the face of death? And does not the other face cause to die in turn, once and for all, everything which operates once and for all? If there is an essential relation with the future, it is because the future is the deployment and explication of the multiple, of the different and of the fortuitous, for themselves and "for all times."[82]

As I read this passage, Deleuze is coming out on Nietzsche's side in favor of pluralism.[83] Let me now see whether there is a similar outcome in *Bergsonism*.

In that book Deleuze focuses Bergson's argument on exactly the place that Merleau-Ponty chose to criticize Bergson. The remarkable difference in styles comes out in Deleuze's summary of the Bergsonian argument that the relativistic physicist has to be, in my terminology, a two-timer:

In order to posit the existence of two times, we are forced to introduce a strange factor: the image that A has of B, while nevertheless knowing that B cannot live in this way. This factor is completely "symbolic"; in other words, it opposes and excludes the lived experience and through it (and only it) is the so-called second time realized. From this Bergson concludes that there exists one Time and one Time only. . . .[84]

The idea is again like that of our imaginary space cadets who cannot really be imagined as living in a different time from that of those who are tied to a planet. Otherwise, we would not be able to say that on their return the cadets were younger than those who stayed on Earth. This claim requires the existence of only one experiential timeline, one framework of temporality from which to assess the two times.

Deleuze is certainly entitled to respond to Bergson's claim that time is one by asking, "One what?" In other words, the metaphysical claim about time needs to be connected to temporality, which is plural. Deleuze tries to capture the difference between Bergson's monism and temporal pluralism by distinguishing two forms of pluralism. "Generalized pluralism" posits an infinity of actual fluxes, whereas "limited pluralism" posits a single "virtual whole" in which all these fluxes participate.[85] The difference between monism and limited pluralism depends on the word "virtual," which Deleuze made famous. The idea is that no one can occupy more than one temporal framework (Kant was right about that), but that framework (the "Whole") is virtual, not actual. "The Whole is not given," says Deleuze of Bergson, but it is "actualized according to divergent lines" that "*do not* form a whole on their own account, and do not resemble what they actualize."[86] Anticipating his own later views about organisms, Deleuze recognizes that this idea of the whole suggests an organism that develops itself toward ends and an ending. He maintains, however, that an organism is not a closed whole unto itself. Instead, the organism opens onto a virtual whole that has finality, because organic life always involves directions, even if there is no final "goal." This teleology without a telos comes about as a result of the directions being created through the actions of the organism. The directionality is different from a goal because a goal is readymade and preexists the actions. The virtual Whole is, then, indicated as a plurality of lines of direction that do not converge on a single point. Deleuze thus provides an interpretation of Bergson that avoids the metaphysical postulate of a single substance and a closed universe. Instead, Bergson is rescued from himself and becomes a philosopher who reconciles unity and multiplicity in a way that transcends the contrast between monism and (generalized) pluralism.

Reflections

This chapter has opened up a series of issues about the past. Here are five of these questions, with the aperçus gained through the foregoing philosophical vignettes.

(1) What is the past? Where does the time go? Common sense would answer that the past is where time goes when it is no longer the present. On Bergson's view, however, it is wrong to think of the past as a "place" where time could go. This way of thinking about time spatializes it. For Bergson what counts as the past is a function of the present. As a cone, the past can be expanded and unpacked, or it can be condensed into the present moment only. Merleau-Ponty calls this way of thinking about temporality "snow-balling." The disadvantage of this way of seeing temporality through the eyes of the present is that it makes the independence of the past difficult to explain. In contrast, a theory such as Husserl's account of retention makes the present always seem to be past.

(2) Can one change the past? As one might expect of philosophers, the answer is yes and no. No, if the past is the context within which our present self-understandings arise and which gives our projects the grid that makes them intelligible in the first place. Yes, in a different sense, namely, if the past is changing as the present does. Interpretations of the past depend on evidence, however, and this evidence anchors the interpretations. Of course, what counts as evidence is itself a function of the interpretation. But that does not mean that the interpretation can take the evidence any way it wishes. Otherwise, the construal of what evidence is would be suspect.

(3) What is memory? For James and Husserl, the sense of time passing is not the same as recollecting a past event in the present. The former is retention, and only the latter is memory, strictly speaking. Memory is an often—but not always—deliberate recall of phenomena that were once present, but are now present only as

past. Retention is a necessary condition of memory, but not a sufficient condition. That is, there could not be memory without retention, but retention does not entail memory.

(4) Is memory continuous? The unity of time is often taken to entail the unity of the smooth passage of retention over time. Even if this were the case, however, the continuity of memory can be doubted. Poststructuralist philosophers are not alone in seeing memory as shot through with gaps and discontinuities. Memory may well be looking back through the lens of an intervening lifetime, and thus the past that it envisions may be a past that never was. Or the sense of continuity may be an illusion generated by the apparent coherence of temporality itself.

(5) But if memory is discontinuous, does that entail that temporality is full of gaps as well? The Husserlian answers by emphasizing retention, whereby the present slides continuously into the past. The contrasting Bergsonian notion, shared by James, is that the past is recollected in the present, and is therefore continuously conditioned by the present (which can change). So even if retention and recollection are not the same (by virtue of point 3 above), the difference between them does not rule out the possibility on both accounts of continuous temporality.

Now it is time to be self-reflective, however. So I take a step back in order to ask about the methodology implied in the act of posing these questions about past philosophers. What issues are raised by the very activity of reading past philosophers in the light of present philosophical interests? Why read past philosophers at all? Why not set out one's own views without all this nostalgia for the past?

To deal with these questions, I identify the method of this book as genealogy rather than as phenomenology. Genealogy is more hermeneutical than descriptive. That is to say, it recognizes itself as an interpretation that is not simply a presuppositionless description, but a view generated by a particular standpoint. Just as there can always be more than one interpretation, furthermore, there can be more than one phenomenological account of human existence. As Heidegger remarks in *The Basic Problems of Phenomenology*, "there is no such thing as *the one* phenomenology."[87] The danger of hermeneutics is that it can forget to apply to itself its own adage that everything is a matter of interpretation. If it forgets to thematize both its connection to the past history of interpretation as well as its own present interests, it will become what it is often accused of being, namely, too tradition-bound, or in a word, *nostalgic*. In this closing section of the discussion of the past, I will indicate briefly what is wrong with nostalgia and how to avoid it by increasing interpretation's self-critical awareness.

A hermeneutical theory of interpretation maintains that there is a circular relation between past and present. Because of this circle, philosophy can never be done without some connection to the past. Let me refer to this as "methodological nostalgia," which is nostalgia in a good sense. This temporal circle is not "vicious" in the logical sense, but is instead a feature to be emphasized and made explicit. If what counts as significant in the past is invariably conditioned by present interests and needs, then these present concerns should be thematized and made as explicit as possible. In particular, the more that an interpretation makes explicit how it has been influenced by the history of preceding interpretations, the less dogmatic it is likely to be. Interpretation should always be open for other possible interpretations. Otherwise, it will overlook the extent to which it is reading its own standpoint into the past and therefore losing sight of what makes the past different from the present.

One example where a philosopher obviously reads his own theory into an earlier philosopher is Heidegger's Kant book. We saw in chapter 1 that in that book Heidegger reads Kant as really saying that time and the I of pure apperception are essentially the same. From our present perspective we can see that Heidegger was reading his own interests in temporality into Kant's concern with time. Heidegger hesitated to say for himself what he made Kant say for him, that is, that original time makes the mind possible and not the reverse. We can say this and believe it at the same time that we recognize that our own temporality is involved. That is to say, our perception that Philosopher A is being misread by Philosopher B does not preclude that we are also misunderstanding Philosopher A, albeit differently than Philosopher B misunderstands Philosopher A.

The point is that the phenomenon of influence works not simply from the past to the present. The present can also influence the past. Or perhaps the lesson of this chapter is that it would be more cautious to say that the present influences the *perception* of the past. But if the past is the way it is only insofar as it is perceived that way, then the contrast between perception and reality is called into question in this particular case.

Thus, we can take some of the themes of this book as cases in point. The perception of Kant as being beyond the realism–idealism controversy, or the account of the contrast between faculty psychology and duration theory, or the discussion of the difference between Husserl's diachronic and Bergson's synchronic explanation of duration are all conditioned by a sense of what the most interesting debates and issues are in *current* philosophy. Those issues will change, of course, and thus later writers will inevitably supersede this book's expectations of what is of interest in these texts. The expectation of a present understanding carrying into the future is not derived from a nostalgic fixation on past problems that one feels one has come to understand and appreciate. Feeling too comfortable with one's sense of what the issues are is nostalgic in a bad sense because it blocks the possibility of new questions coming into being. Nostalgia in the bad sense turns the present into the already past and it ignores the problem of the future leading to a different sense of the present. The future raises the unsettling worry that this present moment has already entered the past. This thought may even suggest that there is no past without the future. To test this thought is the task of the next chapter.

4 "The Times They Are a-Changin' ": On the Future

If nostalgia is one side of the coin, hope is the other side. Nostalgia is putting all one's hope in the past. The previous chapter maintained that nostalgia is to be avoided. Does avoiding nostalgia therefore mean giving up hope for a better future? This is a central question in the politics of temporality. In this chapter, I will consider the advantages and disadvantages of hope. Hope can imply too much continuity with the past, such that total change becomes unlikely. In contrast, hope for total change can blind the politically active to possibilities in the present.

For this debate to make sense, much depends on what is meant by the future, a word that perhaps should always be followed by a question mark. An initial distinction concerns the "future" in both the phenomenological and in the historical senses of the term. This is not a distinction simply between an individual's future and a collective future, although that is certainly a central part of the difference. In the historical domain, there is also the difference between the teleological and the eschatological sense of the future. Although these are normally run together, they are not identical in meaning. Teleology implies an account of the developmental emergence of social and political events and structures. Eschatology, in contrast, suggests a sudden, disruptive occurrence such that when it happens is irrelevant. The eschatological event could happen tomorrow or centuries from now. I will explain below how Kant and Hegel project teleological accounts of the historical future, with Kant's being more eschatological and Hegel's being more teleological. Insofar as the scope of these accounts involves the history of all humankind, it is called "philosophical history" or "universal history," and is more in the domain of the philosopher than the historian.

Not every philosopher shares this interest and even belief in universal history. Schopenhauer and Nietzsche, for instance, receive these ideas with skepticism. Schopenhauer dismisses universal history as seeing shapes in clouds, and Nietzsche sees it as an elitist abuse of history. In a famous parable Walter Benjamin gives a Nietzschean twist to the Marxian dialectics of history. Benjamin's parable of the *Angelus Novus* salvages a minimal hope from the collapse of the ideal of universal history that then has echoes in the recent political writings of both Jacques Derrida and Slavoj Žižek.

If Derrida and Žižek differ over the status of subjectivity and consciousness, they both face the difficult problem of how to justify their skepticism about universal history with their hope for the possibility of progressive politics. Derrida in particular starts by returning to a more phenomenological sense of the future. English cannot capture in its single word, "future," the two senses that Derrida distinguishes in French: *l'avenir* and *le futur*. For Derrida, the latter is the predictable future that is expected, whether for better or worse, whereas the former is the unpredictable, unexpected event, again, for better or worse. For example (not one of Derrida's), we might expect global warming to bring the environment crashing down, and yet we might hope that taking the right ecological measures will avert disaster. That we do not know whether anything we can do will rescue us makes this hope for the unexpected seem empty. To give up hope entirely is to give in to despair, however. So a hope that generates continued ecological efforts can still be a more effective and pragmatic attitude than the cynicism and do-nothingness that result from despair. Derrida thus refers to the future in the sense of the unpredictable as "messianic." He himself wants "messianicity without messianism." Is this the same as "hope without hope"? Can a politics be built around this notion, perhaps much in the way that Slavoj Žižek calls for a "Bartleby politics" of active refusal? Or does it lead to fatalism, empty optimism, and silent despair? Žižek thinks that Heidegger's later politics of *Gelassenheit* does indeed lead to political quietism, and Žižek therefore replies to it with a resounding "No thanks!" The goal of this chapter is to assess the politics and the ethics of the competing philosophical accounts of the future. The first task, however, is to awaken the phenomenology of the futural correctly. Then we can see where we stand on the historical question of whether we should have hopes for a better future, or whether we have to accept that these are posthistorical times in which those hopes are being gradually abandoned.

Kant and Hegel on Universal History

Does hope require a final telos? Without a goal for history, is there no hope? Do we need the ideal society that is invariably in a future-that-never-comes to judge present-day society? In this section I want to review Kant's and Hegel's views about universal history before turning to more recent, and largely negative, answers to these questions.

Kant lists four questions that should be answered by philosophy: What can I know? What ought I to do? What can we hope for? What is man? In the present context the third question is the most pertinent. What we can hope for is not an individual matter, but involves social regulation of needs and desires. I can pursue my own ends only insofar as they do not conflict with other people's needs and desires. Kant addresses these issues in several essays on the philosophy of history, beginning with the essay from 1784 entitled "Idea for a Universal History from a Cosmopolitan Point of View." By "Idea" here he is invoking his technical term. An Idea of Reason is a thought that we can entertain rationally, but for the truth of which we can never have sufficient evidence. God, free will, and an immortal soul are other such Ideas of Reason, unlike the categories of the Understanding, without which we could not have experience. In particular, Universal History involves a regulative Idea, one that can be approached asymptotically but never attained. Universal progress toward social and political freedom is achieved not by reaching the point where there are no longer any constraints on human desires, but by reaching the point where humanity learns to accept constraints. For Kant, the point where freedom and constraint are in balance, which for him is the telos of the history of humanity, is the perfectly just civic constitution. The perfect constitution is the regulative Idea whereby individuals' freedoms would be maximized and conflicts with others' freedoms would be minimized. Kant argues that even if this ideal seems utopian and impossible, we can nevertheless hope that it is where we are headed.

In the essay "Perpetual Peace" (1795) Kant emphasizes this hope by arguing that even a race of devils could be expected to achieve a perfect constitution. Kant maintains that we do not need to know "how to attain the moral improvement of man but only that we should know the *mechanism of nature* in order to use it on men, organizing the *conflict of the hostile intentions* present in a people in such a way that *they must compel themselves to submit to coercive laws*."[1] In other words, it is in everyone's self-interest to restrict his or her own wants. Kant's reasoning is that even the most self-interested creature should realize that it could not get what it wanted unless other such creatures restrained themselves as well.

Whatever our conscious intentions, then, this increasing self-regulation of society runs like a railroad track into the future. In the essay "What Is Enlightenment?" (1784) he claims optimistically that people "work themselves gradually out of barbarity if only intentional artifices are not made to hold them in it."[2] Even if individuals wanted to jump the tracks, they would not be able to because Nature has built us in such a way that we approach the ideal as if we were on rails. Kant is both a pessimist about human nature, which he defines as involving a necessary propensity for evil, and an optimist about human history, since he thinks that history shows that we are necessarily approaching the regulative Idea of the perfect constitution.

Enlightenment means the escape of humanity from its "self-incurred tutelage."[3] "Tutelage" involves the yoke that people come to identify with because the fetters make them who they are. The hope is that because the tutelage is "self-incurred," it can also be left behind. Enlightenment can come about as a result of freedom. "If freedom is granted," says Kant, "enlightenment is almost sure to follow."[4] Note that enlightenment follows from freedom, and not freedom from enlightenment. Kant refers to this as the priority of practical reason over theoretical reason.[5]

Kant recognizes that hope must be based on there being some evidence for the plausibility of this projection of universal history. He therefore gives three reasons for believing in the attainability of this end. The first is the systematic structure of the cosmos, as argued for in the first *Critique*, plus the claim of the third *Critique* that Nature is teleological and does nothing in vain. The second is the claim that we cannot be indifferent to this hope for a universal cosmopolitan condition, even if we do not share it. The third is the universal sympathy that he observed to be felt for the aims of the French Revolution (even if he could not condone the Revolution itself, given his rejection of a right to revolution).

Kant therefore calls this necessary progress toward maximal freedom the *ruse of Nature*. Hegel modifies this notion into the *cunning of Reason*. For Hegel, individuals act for their own personal reasons, but what they accomplish may be more than what they intended. Napoleon may have wanted to aggrandize his own power, for instance, but in the broader historical picture his unification of Europe pointed to the possibility of an encompassing unity of all humankind. Marx takes over this idea when he says that history cannot happen without individuals' willing, even though what actually happens will not be what individuals will.

Reason for Hegel implies that progress in history depends on becoming increasingly conscious of what we want to achieve. For Hegel, consciousness is crucial to becoming free: "No truly ethical existence is possible unless individuals have become fully conscious of their ends."[6] Hegel thereby reverses Kant's thought that enlightenment follows from freedom. For Hegel, without enlightenment freedom could not be obtained. Hegel's view thus stands in contrast to Kant's, which posits that it does not matter what motivates individuals because a cooperative collectivity will emerge even if it is not in our nature to be sociable. For Kant, our unsocial sociability is Nature's ruse for bringing freedom into being, since Nature itself is completely deterministic. Humanity wills concord, remarks Kant, but Nature wills discord. The point is that we can hope for concord to emerge from the discord that we find all around us precisely because humanity will learn to give up its brutish individual desires, even if only out of self-interest, in favor of a lawful constitution.

Hegel has doubts about this leap of faith, and instead wants to provide an account that gives a better explanation than relying on this belief in the providence of Nature. On Hegel's account, we pull ourselves up by our own bootstraps, rather than by running along Nature's rails. Hegel makes this universal history into the story of the gradual emergence in self-consciousness of the ideal of freedom. First one individual is free, then some are free, and then it is universally realized that everyone is free. Each of these levels is a historical achievement. Once an advance is made in self-consciousness, there can be no backsliding. This is not to say that scarcity may not come along and reduce actual freedoms. Conceptually, however, the freedom of everyone, once recognized, necessarily becomes the object of hope. The hope for the progress of freedom is thereby shown to be rational.

Both Kant and Hegel thus paint a utopian picture of universal history for all of humankind as the paradigm of cosmopolitanism. Skeptics about this picture fear that such hope for the future only serves to cover up the injustices of the present. Hope may also serve to cover over the sufferings of people in the past. Frankfurt School critical theory and French poststructuralism have long resisted this enlightenment philosophy inherited from Kant. Let me now consider some alternative phenomenological analyses of the future and review the recent history of resistance to this hope.

Heidegger on the Futural

Whereas Kant and Hegel give us a philosophy of history for the species, they have little to say about the phenomenology of the future. They do not go into what is involved in the individual's sense of the temporality of the time that is ahead, that is yet to come. Heidegger is the first to provide a genuine phenomenological account of the future. Even Husserl had less to say about protention than he did about retention. For Heidegger the temporality of the futural is what allows him to distinguish the authentic from the inauthentic in the second division of *Being and Time*. This distinction is not so much value-laden as it is the basis for distinguishing values as affirmative or negative. Heidegger makes this distinction in his explanation of the ontological source of ontic value discriminations. Let me attend to this distinction before discussing in particular Heidegger's prioritization of the future. Then we will be able to observe subsequent philosophers trying to find their way between the Heideggerian phenomenological account of the priority of the futural and the Kantian-Hegelian historical account of universal progress.

In §68 of *Being and Time* Heidegger explains the difference between authentic and inauthentic temporal comportments. There will be three of these comportments, corresponding to the three "ecstases" of temporality. The term "ecstasis" is coined so that the past, present, and future will not be construed as separate times or zones of time. Instead, each of these is a direction that temporality can take. In addition to past, present, and future, furthermore, Heidegger says that each of the three fundamental existential dimensions of Dasein—situated attunement, projective understanding, and falling (into the present)—is tied more to one of these ecstases than the other two.

Understanding as the projection of possibilities looks ahead toward the future whereas attunement reflects a sense of situatedness and thus of the past. The authentic mode of the understanding involves a sense of the future as a "coming toward" as opposed to the inauthentic understanding of "going-into." The latter represents an attitude of simply awaiting, or merely expecting, things to occur to one. This attitude contrasts with authentic anticipation that, as a *Vorlaufen*, "runs ahead" and seizes the possibilities that are more important while avoiding the distraction of less important matters. The authentic and inauthentic comportments of the past and present understandings are then correlated with this account of the future. The inauthentic past understanding is characterized as "having forgotten" questions of importance and unimportance and even as "backing away" from them. The authentic understanding of the past Heidegger calls (following Kierkegaard) "repetition," which is an explicit avowal and taking on of a possibility. The contrast between an inauthentic "making-present" which then becomes "lost" in everydayness and the authentic "moment of vision" (*Augenblick*) was mentioned previously in the discussion of Heidegger's account of inauthentic understanding, which presents by way of objectifying and isolating possibilities rather than by actively integrating them into a connected life.

These contrasting attitudes of the understanding and its futural projection are reflected in relation to attunement and falling as well. Drawing on Heidegger's examples, an inauthentic attunement vis-à-vis the future may be the emotion of fear, whereby in the face of a particular threat one "comes back" to one's potentiality-for-being.

In contrast, Angst, as we have seen, is an authentic attunement whereby one calmly and "resolutely" faces up to one's finitude. "He who is resolute," says Heidegger, "knows no fear."[7] Though this generalization may be overly heroic, it does bring out the relation to one's past, for Angst brings one back to one's thrownness as a possibility that is still live or real, and that can be repeated. As for the relation of the attunement toward the present, Heidegger says that Angst holds the *Augenblick* "at the ready" ("*auf dem Sprung*").[8] Anxiety discloses the insignificance of the everyday world and brings one back to basic issues and concerns. This resoluteness contrasts with inauthentic or fearful comportment such as seeing the present as "lost" or of forgetting one's past aspirations.

The discussion of "curiosity" is one of Heidegger's clearest examples of falling into the present in a way that looks away from rather than toward the future. Curiosity's relation to the future is to see only in order to see, to have seen, and to be seen. "Curiosity is futural," says Heidegger, "in a way which is altogether inauthentic, and in such a manner, moreover, that it does not await a *possibility*, but, in its craving, just desires such a possibility as something that is actual."[9] In other words, like a tourist traveling from city to city just to have seen the "top ten" sights, curiosity's relation to the past is to distract oneself by "leaping away" and "not-tarrying," to the point of "never-dwelling-anywhere." The relation to the past becomes a jumble of centuries as the tourist jumps from the medieval cathedral to the museum of modern art, and then back to the Renaissance, without any sense for the real duration that was involved in the development of art.

There is obviously a close connection for Heidegger between the past and the future, and one advantage of his notion of the ecstases of temporality is that the ecstases are interconnected to a degree other theories of time may not recognize. Consider, for example (not one of Heidegger's, of course), the section of Disneyland called Tomorrowland. This supposedly "futuristic" part of the theme park was always clearly yesterday's tomorrow. The rides and scenes of Tomorrowland reflected a vision of the future of the 1950s, with the submarine ride, the people mover, and the racing cars. Of course, such a future can be reestablished by an aesthetics of the "retro." The retro posits itself as the future of a particular past, or as a "future past."[10]

In brief, there is no end to the intricate interlacements of future and past. The feature of Heidegger's account of temporality that is especially pertinent for this chapter concerns the priority that he gives to the future. In *Being and Time* he gives the future priority by way of the analysis of being-toward-death. For many years commentators and critics assumed that he attributed priority to the future *because* he privileged death. Now, however, we have access to many previously unpublished materials that have become available since his death. From these we can see that his main concern was with the priority of the future, and not with death per se. What these works show are other routes than being-toward-death for an argument that could establish the sense of temporal direction that Heidegger wants to establish.

The flow of temporality is ordinarily thought to be from out of the past into the present and on to the future. For Heidegger, however, the authentic sense of the flow is that temporality comes from the future into the present by way of the past. As he defines it, "Temporality temporalizes itself as a future which makes present in the process of having been."[11] The future is thus not some "now" that may or may not show up. The future is instead a necessary feature of the present and the past. Whereas the normal way of thinking about the future is as time that is still to come, from the phenomenological point of view on temporality there could not be a present or a past without a future. Even someone who was about to die in the next instant would still have a future. The shortness or length of the future is irrelevant. Provisionally it will be clearer to speak of the *futural* rather than the future to distinguish the phenomenologically futural from the objective future.

Heidegger discusses the futural in terms of a relation to taking over one's thrownness and relating to one's facticity. In *Being and Time*, he says, "But taking over thrownness signifies *being* Dasein authentically *as it already was*. Taking over thrownness, however, is possible only in such a way that the futural Dasein can *be* its ownmost 'as-it-already-was'—that is to say, its 'been' [sein "Gewesen"]."[12] The phrase "taking over thrownness" implies that one must continue to self-identify with one's past. That is to say, one should continue to live in a manner that is consistent with the way one has always lived. We can call this the directive of "appropriating oneself."

This interpretation is only a part of the story, however, for it takes Heidegger to be making primarily an ontic claim rather than an ontological one, where "ontic" means a concern for the parts of our everyday world and "ontological" implies a grasp of the whole as being what we care about. If this were merely an ontic account, there would be no reason why Heidegger could not equally well say that one can decide that one's past life was a total mistake and vow never to repeat it. "Reinventing oneself" in this fashion is also a way of taking over thrownness, and it seems just as good as "appropriating oneself." Heidegger's point is not limited to the question of whether "self-appropriation" or "self-reinvention" is the better strategy. He is also arguing for the ontological claim that however one relates to one's past, whether by appropriating it or reinventing it, there is a necessary connection to the future involved such that one could not have a past unless one had a future:

Only in so far as Dasein *is* as an "I-*am*-as-having-been," can Dasein come towards itself futurally in such a way that it comes *back*. As authentically futural, Dasein *is* authentically as "*having been*." Anticipation of one's uttermost and ownmost possibility is coming back understandingly to one's ownmost "been." Only so far as it is futural can Dasein *be* authentically as having been. The character of "having been" arises, in a certain way, from the future.[13]

The past is not alone in requiring the futural. Even the present, in the form of the authentic moment of vision, necessarily involves a futural projection: "The moment of vision, however, temporalizes itself in quite the opposite manner—in terms of the authentic future."[14] A central claim in Heidegger's phenomenological analysis of temporality is that the futural is a necessary dimension of any sense of the past or the present. Thus, one cannot even speak of the past or the present without implicating the futural. I thus see Heidegger as being closer to Bergson than to Husserl insofar as he is saying that the futural is part of the present, and is not the same as the future nows that are yet to come. The latter would be a mistaken ontic interpretation of Heidegger's insistence on the priority of the future. Instead, Heidegger is making an ontological claim about the necessary involvement of futural projection in the directionality of time.

Walter Benjamin's *Angelus Novus*

In *Being and Time* Heidegger is theorizing our everyday ways of comporting ourselves in the practical world. In that sense, he is offering us a theory of practice and not just a theory of theory. Frankfurt School critical theory also sees itself as a theory of practice in opposition to traditional theory. Although Walter Benjamin was not a member of the Frankfurt School, he can be taken as a fellow traveler of critical theory because of his connection to Adorno, who gave him significant financial support. Unfortunately, unlike Adorno, Benjamin never made it to Los Angeles because he was either a forced suicide or a murder victim. (We do not know if he killed himself because he could not escape into Spain, or whether he was murdered in attempting to do so.) He did leave us a powerful image of the angel of modern times in his interpretation of a small painting, *Angelus Novus* by Paul Klee, which Benjamin owned. In chapter 3 we saw Pierre Bourdieu undermining Heidegger's argument for the priority of the *temporally* futural.

Now I will examine how Walter Benjamin's analysis of Klee's painting undercuts the priority that Kant, Hegel, and Heidegger give to the *historically* futural. That will prepare us for Derrida's critique of Benjamin and for the development of Derrida's notion of the future to come as messianicity without messianism.

Benjamin's interpretation of Klee's painting in "Theses on the Philosophy of History" (1936) resembles Heidegger's account of time as coming from the future, except for one feature: the angel is going into the future facing backward. Thesis 9 is worth quoting in its entirety:

A Klee painting named "Angelus Novus" shows an angel looking as though he is about to move away from something he is fixedly contemplating. His eyes are staring, his mouth is open, his wings are spread. This is how one pictures the angel of history. His face is turned toward the past. Where we perceive a chain of events, he sees one single catastrophe which keeps piling wreckage upon wreckage and hurls it in front of his feet. The angel would like to stay, awaken the dead, and make whole what has been smashed. But a storm is blowing from Paradise; it has got caught in his wings with such violence that the angel can no longer close them. This storm irresistibly propels him into the future to which his back is turned, while the pile of debris before him grows skyward. This storm is what we call progress.[15]

Insofar as the angel's back is to the future, Benjamin's suggestion is that our historical temporality is really more backward-looking than forward-looking. What Benjamin's parable brings out is the unconscious tendency of universal histories such as Kant's and Hegel's to assume that we are going into the future facing forward. Universal progress is achieved through a valorization of forward-looking visions. In contrast, Benjamin wants to emphasize the backward-looking orientation of critical theory. "Backward-looking" does not imply that Benjamin's position is reactionary. To be antiprogressive is not necessarily to be regressive. On the contrary, Benjamin's critical attitude derives from thinking that forward-looking, utopian visions often overlook massive injustice in the past and present. When specifically contrasted with Heidegger's account of "projection" as consciously and resolutely positing a telos, Benjamin's critique is that what we really see is not purpose and meaning in our lives, but contingency and confabulation. As the angel is propelled into history, it looks back. In looking back, the angel does not see the connected and sequential chain of events that a forward-looking agent would envision. Instead, Benjamin's angel sees one single catastrophe, the wreckage of which accumulates at his feet. For Benjamin, the time of our lives does not get progressively better, but when viewed backward, it appears to be disjointed, out of joint, discontinuous, a series of fragments.

I want to raise and answer four questions about Benjamin's allegory. First, is there not some tension between the directionality of the storm, which is blowing the angel into the future, and the pile of debris, which builds up at his feet, growing skyward, as if he were not moving away from each bit of wreckage? The answer requires us to consider the temporality of the debris. Presumably the debris is not left behind as the angel is blown into the future. Instead, the debris goes along with the angel, piling up at his feet. The significance of this point will become clear as the other questions are answered.

The second question is, given that the meaning of history has crumbled into a pile of debris, what gives Benjamin the right to speak of "one single" catastrophe? The answer is that what is one and single is not the debris, but what it is that has been wrecked, namely, universal history and the very idea of global progress. Kant and Hegel see history as the story of the development of universal reason and freedom. In contrast, Benjamin's angel sees that this story of the progress of civilization is an ideological shambles that distorts and enervates the present. We are at the mercy of the storm, and the message is that our sense of ourselves as moving forward is an ideology that ignores the victims of history and the reality of barbarism.

The third question concerns the directionality of time, and our own temporal orientation. Unlike the famous critique of Hegel by Marx, Benjamin neither stands us on our heads nor puts us back on our feet. Instead, he turns us around so that we are facing backward. The point of the angel's facing backward is that history has no telos. Unlike Marx's spatial metaphor, which has the Hegelian seeing the world upside down, Benjamin's temporal metaphor implies that we cannot see where we are going. Are we in fact going backward? No, because we are moving away from where we have been, not back to where we were before. The story is still linear. However, it is difficult to say that we are moving forward. There are no signposts, no indications of an increase in freedom. The wreckage just piles up and apparently leaves no basis for teleology. The debris does not contain the continuity and coherence of a narrative that would allow us to think of ourselves as approaching a telos. "There is no document of civilization," says Benjamin in the seventh thesis, "which is not at the same time a document of barbarism."[16]

Does Benjamin thus deprive us of any sense of the time of our lives, a sense for how a life is connected between birth and death? Benjamin suggests that past generations provide the present with a "*weak* Messianic power."[17] What is the basis for this messianic hope, given the starkness of the figure of universal wreckage? The answer is similar to the account that I just gave of the temporality of the debris of universal history. As the storm blows the angel backward, the debris is not strewn out in the receding distance, but accompanies him, piling up at his feet. The present is not "empty," homogeneous time, but rather is what Benjamin calls *Jetztzeit*, the momentous moment with the potential to change the direction of history: "The present, which, as a model of Messianic time, comprises the entire history of mankind in an enormous abridgment, coincides exactly with the stature which the history of mankind has in the universe."[18] The messianic moment does not come from knowing where we are going, but from seeing where we have been.

Citing Nietzsche, Benjamin says that the image of enslaved ancestors provides more motivation than that of liberated grandchildren.[19] This vision of past enslavements is not the beginning of knowledge of how things could be better, although it does lead to the knowledge that universal, progressive history is untenable.[20] There is, after all, no standpoint from which to observe the entirety of history. Universal history is written from outside or at the end of history. But we are always only ever *in* history, and its end is always in a future—one that will never come.

For Benjamin the past becomes critically significant in the moments of great social danger. In thesis 6 of the "Theses on the Philosophy of History," Benjamin writes,

> To articulate the past historically does not mean to recognize it "the way it really was" (Ranke). It means *to seize hold of a memory as it flashes up at a moment of danger.* Historical materialism wishes to retain that image of the past which unexpectedly appears to man singled out by history at a moment of danger. The danger affects both the content of the tradition and its receivers. The same threat hangs over both: that of becoming a tool of the ruling classes. In every era the attempt must be made anew to wrest tradition away from a conformism that is about to overpower it.[21]

Memory is like a shooting star. It must be seized hold of, or memorialized, the instant that it flashes by. Insofar as Benjamin defends the idea of a history of the victims, Benjaminian historiography brushes history against the grain. Perhaps it even changes the past, although not in the deliberate if arbitrary way that it happened in the old Soviet Union with each change of leadership. Or if talk of changing the past is too unrealistic, then we can say that what changes is our understanding of the past.

The fourth question to raise about this parable is, what is the wind? What tears us out of a past that perhaps never existed and thrusts us toward a future that probably will never come? If my analysis is on the right track so far, one answer that suggests itself is that the wind is temporality as such. The wind's strength indicates not simply the *flow* of time but the *force* of time. Temporality, or time as experienced, is directional even if it has no particular direction or telos. We can see what is behind us even as we are forced to leave it behind. The wind's force also points to its irreversibility. There is no going back. The reason for this is not Hegel's optimistic assumption that once the ideal of universal freedom appears, there can be no conceptual backsliding. For Hegel, once the ideal becomes conscious, it cannot be forgotten, even if past and present social arrangements fall far short of the future society that it envisions. For Benjamin, in contrast, what prevents us from going back is the fact that the past is so atrocious. Hopes for progress toward peace and prosperity need to be critically unmasked by revealing the underlying economic inequalities that led to the massive wars and systematic slaughter of millions of people in the twentieth century.

Benjamin's parable thus tells the story of temporality as having directionality even if no direction. Temporality is also shown to have irreversibility in the sense of "going away from" even if there is nothing that it is going toward. Benjaminian temporality thus has force even if it cannot be said to "flow." The political implication of this analysis of temporality is clear. Universal history, which tells the story of the continuum of progress toward universal freedom, must be abandoned because it empties human freedom of concrete content. Universal history also leads to fatalism insofar as progress occurs automatically, whether mechanistically in a linear direction (Kant) or dialectically in a spiral one (Hegel).[22] The point of writing history against the grain is not to prove that this continuum obtains, but to blast it open.[23]

To conclude this discussion of Benjamin, I will point out that he prefers the temporality of the calendar to that of the clock. The reason is that instead of the clock's smooth flow of uninterrupted transition, the calendar suggests a more punctuated sense of time. The calendar permits the recurrence of days of remembrance. The sense of time conveyed by a clock is continuous transition, whereas the calendar allows for a sense of time as coming to a stop and standing still. The calendar marks the possibility of the day when class action can explode the continuum of history.[24]

Deleuze on the Temporality of the Self

To prepare for subsequent discussion of the disappearance of teleology in this history of the "future," we need to understand the Nietzschean views of the self and time as transmitted into the French tradition by Gilles Deleuze, particularly in his 1968 classic, *Difference and Repetition*. Throughout the earlier discussion of Bergson and Deleuze on temporality, little was said about the self. To discuss Deleuze's view of the future, I need first to summarize Deleuze's analysis of subjectivity. The main question concerns whether there is a self from the beginning of temporality, or whether it comes into being only later in the process of temporalization. Deleuze follows Condillac and Hume in maintaining that the foundation from which the living present and all other psychic phenomena derive is *habit*. Deleuze joins with Kant and Nietzsche in the hypothesis of the modularity of the mind:

> Underneath the self which acts are little selves which contemplate and which render possible both the action and the active subject. We speak of our "self" only in virtue of these thousands of little witnesses which contemplate within us: *it is always a third party who says "me."* These contemplative souls must be assigned even to the rat in the labyrinth and to each muscle of the rat.[25]

Following Nietzsche's lead, Deleuze makes much of these thousands of habits, and of how the self is fashioned by them rather than being some preexistent thing to which the habits accrue. For Nietzscheans the self is a product of underlying modular subprocessors, not a generative agent. Agency is the double of a contemplative self that surveys the thousands of interactions required to integrate tiny actions within a more complex apparent action.

The self and the subject are not the same as the "me." If you do not recognize yourself in Deleuze's account of who you are, that is because for the most part, when we talk about who we are, we have in mind the empirical "me," not the transcendental "I." The "me," moreover, is for Deleuze the result of an interpellation by a third party. As is indicated by the line "it is always a third party who says 'me,'" for Deleuze the "me" is always an other to the "I." The "me" that you think you are is thus not the same as the self that you take to be the agent of your actions.

"Selves," remarks Deleuze, "are larval subjects."[26] He then clarifies this point by saying, "The self does not undergo modifications, it is itself a modification."[27] In other words, the self is not primordial; it is not there all along. Instead, it is a developing structure that can add rules and other features, until it emerges as it is. Or rather, Deleuze prefers to say, "one is only what one *has*."[28] In Deleuze's language, Being, or the way the self *is*, comes as a result of Becoming, of what the self *has*, that is, habits that it has acquired:

These thousands of habits of which we are composed—these contractions, contemplations, pretensions, presumptions, satisfactions, fatigues; these variable presents—thus form the basic domain of passive syntheses. The passive self is not defined simply by receptivity—that is, by means of the capacity to experience sensations—but by virtue of the contractile contemplation which constitutes the organism itself before it constitutes the sensations. This self, therefore, is by no means simple: it is not enough to relativize or pluralize the self, all the while retaining for it a simple attenuated form.[29]

In the terminology of both analytic and genealogical philosophy Deleuze's larval subject could be said to be *emergent*.[30] The subject does not exist fully developed from the start, either structurally or concretely. The most that one can say is that if the subject is there at all from the beginning of experience, it is (to use Bergson's term) so "contracted" into a point that it is barely perceptible. As it unfolds and matures, the larval subject creates through Repetition. In other words, to produce something new there has to be a contrast with something that is not new. Deleuze thus says we produce something new only on the condition that we repeat—once in the mode which constitutes the past, and once more in the present of metamorphosis. Moreover, what is produced, the absolutely new itself, is in turn nothing but repetition . . . , the repetition of the future as eternal return.[31]

Deleuze's invocation of Nietzsche's notion of eternal return within a Bergsonian context is an example of how something creative, original, and new can emerge from preexisting elements that are repeated, but with a difference. To be creative or original, affirms Deleuze, requires identifying oneself with figures from the past.[32] Deleuze's conjoining of Bergson and Nietzsche produces a novel interpretation not only of duration and eternal return, but also of the future.

Deleuze develops this conception of the future as the third synthesis of temporality. His conception of temporality is compatible with his notion of the self as a multiplicity of competing elements. As a multiplicity, the Deleuzian self differs from the Cartesian cogito, which is reduced to an instant, and which exists only through God's continuous creation of succeeding instants. Descartes has thus, says Deleuze, *expelled time*.[33] Kant is then the next step in this secularization of time. Kantian transcendental philosophy represents for Deleuze the death of God insofar as Kant's "greatest initiative . . . was to introduce the form of time into thought as such."[34] When the mind becomes the source of time, there is no need for God any longer.

Once time is thoroughly secularized, temporality becomes visible. Temporality involves three "syntheses," which reinvent Kant's three syntheses or "dimensions" of time described above in chapter 1.[35] The Humean first synthesis is through habit, and it generates the living present as a foundation for the past and future. Memory is then the Bergsonian second synthesis, which is the pure past and which causes the passing of any given present and the arrival of another present. The third and final Nietzschean synthesis constitutes the future, which Deleuze calls the "royal repetition."[36]

Whereas habit is the *foundation* of temporality, and memory is the *ground* of temporality that causes the present to pass into the past,[37] the synthesis of the future is the *order* of temporality, and it generates "the totality of the series and the final end of time."[38] This third synthesis is perhaps the most significant insofar as "it unites all the dimensions of past, present, and future, and causes them to be played out in the pure form."[39] The future is also the source of Deleuzian multiplicity. "If there is an essential relation with the future," Deleuze remarks, "it is because the future is the deployment and explication of the multiple, of the different and of the fortuitous, for themselves and 'for all times.' "[40]

Deleuze's future is thus not so much a dimension of temporality as that which constitutes the difference between the other temporal dimensions, the past and present. To understand this point we should return to his subtle and original account of Nietzsche's idea of the eternal return. Of course, Deleuze's reading is not an interpretation that Nietzsche himself could have given insofar as it depends on Deleuze's Bergsonian account of temporality. As a philosophical reconstruction of what eternal return could mean within a Deleuzian framework, however, it stands apart. One sentence in particular from *Difference and Repetition* sums up Deleuze's interpretation: "The subject of the eternal return is not the same but the different, not the similar but the dissimilar, not the one but the many, not necessity but chance."[41] This account is worked out in more detail not only in *Difference and Repetition*, but also in his book on Nietzsche, and in a short summary in *Pure Immanence.* Deleuze applies his notions of the "virtual" and "simulacra" to this notion. Simulacra undermine the Platonic distinction between the original and the copy. Differential terms, or binaries, are possible only as systems that are themselves simulacra. These systems produce the differentiations that first allow items to be compared on the basis of resemblance. In short, what this comes down to is that "the same and the similar are fictions engendered by the eternal return."[42]

Deleuze does not mention here, although he was certainly aware of it, Nietzsche's early unpublished essay "Truth and Lie in the Extramoral Sense," where Nietzsche says that we falsify experience by perceiving sameness rather than difference. Metaphor, for instance, is useful for survival because it allows us to overlook all the differences in what we perceive in order to pick out objects that resemble one another. We transform Becoming, which emphasizes difference, into Being, which fixes multiplicities into identity. This is the full reason why, as I maintained previously, Deleuze reads Nietzsche's eternal return, not as the return of the Same—which is just one way that Nietzsche sometimes has Zarathustra state the doctrine—but as the selective return of affirmative repetition. Rather than every little detail returning, for Deleuze only things that are affirmed recur. Laziness, for instance, if it returns, returns as something different, if only because one has said "yes" to it, or affirmed it. Nietzsche thus represents the affirmation of difference rather than the identical, of the multiple rather than the One, and of temporality rather than time.

There is an important methodological consequence of Deleuze's account of difference and repetition where I see him to be forging links with the pragmatist and deconstructionist criticisms of metaphysical binaries. Given his analysis of repetition, note that it becomes impossible to say of any two dualistic terms (such as mind and body, male or female, individual or society, public or private, time or temporality) which is primordial and which is derived. Deleuze says astutely,

Repetition is no more secondary in relation to a supposed ultimate or originary fixed term than disguise is secondary in relation to repetition. For if the two presents, the former and the present one, form two series which coexist in the function of the virtual object which is displaced in them and in relation to itself, *neither of these two series can any longer be designated as the original or the derived.*[43]

With this move Deleuze removes himself from the neo-Kantian attempt to find conditions of the possibility of experience. There is no need to discuss questions such as which came first, the chicken or the egg. The reason for this is neither simply because the right answer is the egg, nor because the question confuses logical and temporal priority. Instead, there is no issue of priority because there could not be one without the other. Transcendental arguments thus become unnecessary, given this deconstruction of metaphysical distinctions.

Derrida on Democracy-to-Come

If Deleuze represents one way of reading Nietzsche to get beyond the political alternative of either hope or nostalgia, Jacques Derrida is another way of appropriating and applying Nietzsche. Derrida's reading is marked, however, by his ambivalence toward the influence of Walter Benjamin. Is the effect of Benjamin's parable of the *Angelus Novus* to make the present abandon all hope for a better future? Or does it display at least a *weak* utopian hope? Although more dystopian than utopian, the parable implies that there is at least some teleology in history, and therefore some grounds for hope. In a sense, hope works backward rather than forward insofar as what we hope for is not so much our own redemption from time as the redemption of past injustices to others through memoralization.

Jacques Derrida rejects even this slight vestige of what he calls "messianism" in history. He remarks, "This text, like many others by Benjamin, is still too Heideggerian, too messianico-marxist or archeo-eschatological for me."[44] Derrida is not a philosopher of history, but he does have an account of the future. Earlier I mentioned his distinction between two different senses of the future.[45] The predictable, foreseeable future, *le futur*, is contrasted with the unpredictable, unexpected future, *l'avenir* that can break into the present unexpectedly at any moment. This analysis separates the teleological from the eschatological in historical time. Derrida was always suspicious of the Kantian and Hegelian stories of universal history. What he wants is a messianicity "without mes-sianism."[46] That is, he does not posit an actual Messiah. The Messiah will never come, because it is the essence of the Messiah to be always about to-come (*à venir*). Derrida's joke is that even running into the Messiah on the street would not prove that the Messiah had finally come. Instead, it would only prove that the particular individual was not (yet) the Messiah. Messianicity is thus the eschatological possibility of an unpredictable, unexpected event that could break into the present at any instant. Derrida thinks that there is still some value in this vestigial bit of eschatology. What he rejects is messianism, which is based on the teleological draw of some remote future ideal. Such future ideals are often only pro-jections of current cultural paradigms into an indeterminate future where the present paradigm is unlikely to be relevant.

How sharp is this distinction between the two different senses of the future? Derrida has argued against John Searle about the nature of distinctions. Derrida maintains that if there is a distinc-tion, there must be a sharp conceptual line between the two terms that are distinguished. Deconstruction works, for Derrida, by iden-tifying vagueness in the concepts that blurs the line and collapses the distinction. Searle disagrees. He maintains that there are many distinctions that are not clean cut but that are still useful. His example is the front of the head and the back of the head. We can make this distinction usefully even if we would not know where to draw the line to separate the two regions.

What this debate brings out is that precision is not always possi-ble or necessary. Imagine a philosopher asking Shakespeare whether his line "Tomorrow and tomorrow and tomorrow" means tomor-row, the next day, and the day after, or whether it means the same day going on endlessly. Demanding subscripts for the sake of preci-sion—as in "*tomorrow$_1$ and tomorrow$_2$ and tomorrow$_3$*"—misses the poetic point, and it certainly destroys the aesthetic effect.

What is the point, then? Saying that the future always brings about the unexpected is not a new message. The future generally turns out differently from what one expects. I take it that Derrida was a subtler philosopher who would not simply let a truism slip into his theorization of the temporal. What Derrida is really pointing to, even if he does not put it this way, is the distinction between historicity and temporality. Temporality and historicity are not the same, even if they are connected. Temporal phenomena are not necessarily historical phenomena. Bergson's waiting for the sugar to dissolve or Husserl's listening to a melody are temporal phenomena, but not historical ones. Thus, Benjamin's deconstruction of the idea of universal history is different in scope from the phenomenological concerns with temporality. Even if there can be temporality without historicity, there can be no historicity without temporality.

The question then arises, how does an account of temporality condition an account of historical experience? Derrida's analysis of messianicity is a good case in point. Messianicity as Derrida uses the term is not tied to a conception of universal history, and it does not rely on notions such as progress, or decline, or cycles. Instead, as a basic condition of temporality temporalizing, messianicity is prior to the whole enterprise of the philosophy of history. Messianicity is built into temporality, and temporality is a condition of the possibility of history. By attaching messianicity to temporality rather than to historicity, Derrida contests any attribution of utopianism to him.

A criticism that is often raised against Derrida and deconstruction is the charge of quietism. Can Derrida's theory give us any reason for action? Can it give us any hope? Or is it a form of resignation, or even a refusal to act? Derrida's answer depends on a discussion of Heidegger's infamous statement in the *Der Spiegel* interview, "Only a god can save us now." This interview dates from 1966, but was not published until 1976, after Heidegger's death. Derrida ties his notion of the messianic to the interpretation of this phrase. Even if messianicity without messianism does not entail a hope for salvation, it does express for Derrida a most basic feature of human temporality. We are by nature messianic, Derrida insists, insofar as we cannot *not* be. The messianic character of temporality follows "because we exist in a state of expecting something to happen. Even if we are in a state of hopelessness, a sense of expectation is an integral part of our relationship to time."[47]

Derrida explains this point at greater length in *Rogues.* The intended effect of his skillful *explication de texte* there is to show that Heidegger's utterance is more complex than it might otherwise seem. "Trust me," says Derrida when he claims to know everything ever written about this interview. In fact, however, Heidegger's complexity comes through as evasiveness on Heidegger's part. Derrida begins by acknowledging his ambivalence about Heidegger, and he notes that Heidegger is one in whom "we have never suspected the slightest hint of democratism."[48] Taking each word in the sentence at a time, he points out that Heidegger says "*a* god," and thus, neither "God," nor "*the* God," nor "the gods." Nor, Derrida notes, does Heidegger say "the *last* god." The last god is mentioned by Heidegger in his *Beiträge.* The last god is not the end of history so much as the vision of another beginning to an "immeasurably" different history. This is a god of the future, of a different direction entirely, not the god of the past, with its projected future of an end to all things. The end might not even come into question in this new beginning. In fact, it could not even be said that this was a new beginning for *us*, since whomever's history that would be, it would not be "ours." The god in the interview from *Der Spiegel* is going to save "us." Hence, it is a different god from the last god of that completely "other" beginning.

To prevent theological misunderstandings, I am going to call the "god" of the interview a "cultural paradigm."[49] A cultural paradigm is what is at stake in cultural politics. A cultural paradigm may not be fully articulated, but it is a matter of intense concern. Heidegger is not willing either to affirm or to deny that the cultural paradigm of the future will be democracy: "I am not convinced that it is democracy [*Ich bin nicht überzeugt dass es die Demokratie ist*]."[50]

In response to the journalists' demands that he talk about the "timely" aspects of democracy, Heidegger answers, in a way that Derrida says is "measured and cautious" but that I view as temporizing, "We must first clarify what you mean by 'timely,' that is, what 'time' means here. [*Zunächst wäre zu klären, was sie hier mit 'zeitgemäss' meinen, was hier 'Zeit' bedeutet.*]"[51]

Whether to welcome Heidegger's proffered cultural paradigm is unclear, and this unclarity is precisely the problem. Can a cultural paradigm that is so empty even be anticipated? What would we resolve ourselves for? What is there to be done? Perhaps a philosopher should not be expected to answer these practical questions. Perhaps philosophy can contribute only at the general level of debate about competing cultural paradigms. Nevertheless, to ask such practical questions is both legitimate and necessary.

If Heidegger does not answer these questions, does Derrida do any better? In *Rogues* he reviews different places where he had previously discussed the idea of democracy-to-come. If *Du droit à la philosophie* (1989–90) is the first place it comes up, the *Force of Law* essay, which was given initially at a conference that same year, features the notion more centrally in its deconstruction of Walter Benjamin's weak messianic hopes. As Derrida says, Benjamin's hope is weak because "there is not yet any democracy worthy of this name. Democracy *remains* to come: to engender or to regenerate."[52] Then in *Sauf le nom* (1993) Derrida makes the important comment that democracy-to-come is not a regulative Idea in the Kantian sense, that is, it is not an ideal that one can approach asymptotically (without ever reaching). Nevertheless, it remains as an inherited promise, "for lack of anything better."[53] In other words, just as democracy is said to be the best form of government if only for lack of anything better, so too is the notion of a regulative Idea the best way to understand it, for lack of any better alternative.

Derrida's reservations about the regulative Idea are threefold. First, it seems like an ideal possibility that is infinitely deferred. In contrast, Derrida suggests in the essay "The University without Condition" (2001) that democracy-to-come is not ideal, but real. It is a genuine demand by the Levinasian other, older than I, on me, "like the irreducible and nonappropriable *différance* of the other."[54]

The second objection is that the regulative Idea sounds like a Kantian rule. Derrida holds a common but controversial view that moral rules are like machines that take away one's decision-making power and thus deny the very autonomy that the regulative Idea idealizes:

In the second place, then, the responsibility of what remains to be decided or done (in actuality) cannot consist in following, applying, or carrying out a norm or rule. Wherever I have at my disposal a determinable rule, I know what must be done, and as soon as such knowledge dictates the law, action follows knowledge as a calculable consequence: one *knows* what path to take, one no longer hesitates. The decision then no longer decides anything but is made in advance and is thus in advance annulled. It is simply deployed, without delay, presently, with the automatism attributed to machines. There is no longer any place for justice or responsibility (whether juridical, political, or ethical).[55]

The reason why I say that this characterization of rules is controversial is that Kantians have argued effectively that it involves a typical but incorrect characterization of imperatives.[56] Whether to follow the rule is not only up to me (insofar as I could perfectly well decide not to), the rule specifies what responsibility is, and without that knowledge, I could not act responsibly.

The third reason is not well articulated, but comes down, I believe, to the apparent lack of evidence for saying that a real society is approaching the ideal, however asymptotically. As I pointed out in my discussion of Kant, we could never have enough evidence that we had achieved the regulative ideal. Indeed, we could now in fact be in the best possible society and not know it (although current affairs suggest that this is highly unlikely). The popular idea that we are at the end of history right now might depend on some such feature of regulative ideals. But because there is much solid evidence that present society is far from perfect, the question is whether we need the regulative Idea to know that we are still far from the ideal society. In other words, there are two claims being made. One is that current society falls short of the ideal. That claim we know is certainly true. The other is that we are approaching the regulative Idea asymptotically. This claim could then be used to allege that one's present society is superior to all others, past or present. That claim is certainly problematic, and moreover, dangerous. It causes us to overlook and thus to perpetuate present injustice in the name of false assumptions about progress toward an ideal end.

Those are some of the reasons why Derrida tends to avoid appealing to regulative Ideas like democracy, progress, and the like. He insists instead "on the absolute and unconditional urgency of the *here and now* that does not wait and on the structure of the promise, a promise that is kept in memory, that is handed down [*léguée*], inherited, claimed and taken up [*alléguée*]."[57] He defines the "to-come" as "not something that is certain to happen tomorrow, not the democracy (national or international, state or trans-state) of the *future,* but a democracy that must have the structure of a promise—and thus the memory of that which carries the future, the to-come, here and now."[58] His intention is thus to avoid the quietism, the inability to act, that is often attributed to deconstruction generally. To determine whether he is successful we will have to go further into the notion of democracy-to-come.

There are five points that Derrida wants to emphasize about his notion of democracy-to-come. First, the term "democracy-to-come" is to be used to *criticize* present democracies for involving and especially for covering up existent suffering, injustice, and inequities. That does not mean that the ideal democracy can become real. The contradictory or "aporetic" character of the ideal democracy prevents its own realization. Force that is not force, respecting singularity at the same time as calling for universal equality, reconciling the social and the individual as well as the public and the private: these apparent impossibilities lead to thinking of democracy-to-come as "an empty name, a despairing messianicity or a messianicity in despair."[59] Admitting that democracy-to-come "will never exist, in the sense of a present existence: not because it will be deferred but because it will always remain aporetic in its structure" does not lead to despair, but instead to "active and interminable critique."[60] The ideal democracy is therefore not an idea that is fixed once and for all, as it is in Kant and perhaps in Hegel, but is instead said to have "absolute and intrinsic historicity."[61]

The second point is that the democracy-to-come cannot serve as a telos of history in the way that it does in the Kantian philosophy of history. It must *not*, therefore, be construed in a teleological way:

Democracy is the only system, the only constitutional paradigm, in which, in principle, one has or assumes the right to criticize everything publicly, including the idea of democracy, its concept, its history, and its name. Including the idea of the constitutional paradigm and the absolute authority of law. It is thus the only paradigm that is universalizable, whence its chance and its fragility. But in order for this historicity—unique among all political systems—to be complete, it must be freed not only from the Idea in the Kantian sense but from all teleology, all onto-theo-teleology.[62]

The idea of global progress thus goes by the board because there is nothing to which it can be applied.

The question is, however, where is this ideal situated? We might think that we could reasonably inquire as to when it might occur. Derrida insists, however, on the unforeseeability of the "to-come," which is "beyond the future."[63] This "beyond" is the third feature that he wants to point out in the notion of the democracy-to-come. "Beyond nation-state sovereignty, beyond citizenship," the creation of a new juridico-political space that "never stops innovating and inventing new distributions and forms of sharing, new divisions of sovereignty" is imaginable.[64]

The fourth feature concerns the close connection between law and justice first discussed in "Force of Law" and then spelled out both in *Specters of Marx* and in his reply to critics of that book, "Marx and Sons." Justice is undeconstructible, even if it must be embedded in a system of law. Every system of law will, then, be deconstructible by virtue of justice. In *Politics of Friendship* this analysis leads to what Derrida calls the question of the name: in the name of what can social criticism be applied today?

Is it still *in the name of democracy* that one will attempt to criticize such and such a determination of democracy or aristo-democracy? Or, more radically . . . —is it still in the name of democracy, of a democracy to come, that one will attempt to deconstruct a concept, all the predicates associated with the massively dominant concept of democracy . . .? What remains or still resists in the deconstructed (or deconstructible) concept of democracy which guides us endlessly?[65]

What "democracy" means depends on the historical context of the day. Thus, keeping the Greek name, "democracy," is itself not simply a rhetorical but also a political strategy. As a political strategy it is indeed legitimate because democracy itself guarantees the right to criticism, including the right to deconstruction. Derrida thus remarks, "no deconstruction without democracy, no democracy without deconstruction."[66]

If democracy can thus be construed as "deconstructive self-delimitation," then the idea of the future must not mislead us into deferring the urgency of action in the present. The fifth point to recognize is that Derrida's notion of messianicity without messianism emphasizes the singular urgency of the present need to challenge or "delimit" itself. "In invoking a *here and now* that does not await an indefinitely remote future assigned by some regulative Idea," Derrida writes, "one is not necessarily pointing to the future of a democracy that is going to come or that must come or even a democracy that *is* the future."[67] Although "democracy-to-come" is not a sentence, Derrida maintains that it is both a constative and a performative. "Democracy-to-come" is a constative insofar as it asserts messianicity without messianism, and it is a performative insofar as one believes in the notion of democracy and answers its call for action in the present. The idea of democracy-to-come is not simply to sit back and wait for democracy to show up. Heidegger was right to identify the temporalization of "simply waiting" as inauthentic. Derrida says,

> the democratic injunction does not consist in putting off until later or in letting itself be governed, reassured, pacified, or consoled by some ideal or regulative Idea. It is signaled in the urgency and imminence of an *à-venir*, a to-come, the *à* of the à-venir, the *to* of the to-come, inflecting or turning into an injunction as well as into messianic waiting the *a* of a *différance* in disjunction.[68]

Then with a surprising invocation of the Bergsonian terminology of duration and contraction he adds, "Finally, and especially, however one understands *cratic* sovereignty, it has appeared as a stigmatic indivisibility that always *contracts duration into the timeless instant of the exceptional decision. Sovereignty neither gives nor gives itself the time; it does not take time.*"[69] Does this suggestion of a domain "beyond time" bring back the Kantian regulative Idea that Derrida wanted to avoid? Better to say, I would have thought, that the future is an ecstasis of the present, and this in itself transports us to the future perfect, when it will have been the case that what is now present to us is the past of a future present. That formulation suggests the temporalization that Derrida is looking for, without positing a Kantian noumenal realm that is beyond time, or even timeless.

Žižek on Bartleby Politics

Many recent European philosophers have been greatly impressed by Herman Melville's story, "Bartleby, the Scrivener." In the story Bartleby gradually withdraws more and more from social interaction. When asked to do anything, he responds invariably, "I would prefer not to." This image is powerful even if Bartleby himself comes to a tragic end. Whereas in the United States the story is generally taken to signify anomie and social indifference of the sort that drives Bartleby's bourgeois associates mad with frustration, for the Europeans it signifies a form of passive aggression that challenges all social codes and civic duties. Perhaps Bartleby is the nostalgic incarnation of the spirit of May '68, and expresses a deeper anarchism that turns into "cynical reason."[70] Or maybe its appeal to Deleuze, Derrida, Negri, and Žižek is that it represents resistance without either nostalgia or hope. Let me turn to the question of what a politics inspired by Bartleby would look like, and how it would contrast, say, with a Heideggerian politics of *Gelassenheit*. The current political scene includes the striking contrast between the qualified call by Derrida for the democracy of the future and the more cynical attitude toward democracy of Slavoj Žižek. The striking difference between them is encapsulated by their readings of Bartleby as the basis for a projective politics of the future.

Before going into detail about Žižek's understanding of Bartleby, I should refer first to the accounts of Derrida and Deleuze, as I find the contrasts between their readings of this story and Žižek's to be revealing of the different attitudes toward the future of democracy. Although poststructuralism is often viewed as apolitical or antipolitical, in fact it is not. On the contrary, the birth of poststructuralism in the '60s renders it more forward-looking perhaps than the cynicism of the first decade of the current century, as exemplified by Žižek.

In *The Gift of Death*, Derrida expresses his admiration for Melville's character, who

responds without responding, speaks without saying anything either true or false, says nothing determinate that would be equivalent to a statement, a promise or a lie, in the same way Bartleby's "I would prefer not to" takes on the responsibility of a response without response. *It evokes the future without either predicting or promising; it utters nothing fixed, determinable, positive, or negative.* The modality of this repeated utterance that says nothing, promises nothing, *neither refuses nor accepts anything,* the tense of this singularly insignificant statement reminds one of a non-language or a secret language.[71]

Notice that Derrida does not attribute a strategy of refusal to Bartleby. Derrida says that Bartleby "*neither refuses* nor accepts anything." Derrida compares Bartleby to Job and to Abraham, thereby invoking Kierkegaard's discussion of Abraham's silence as he carries out the commandment to kill his son. The contrast between religious belief and secular society no longer features in Melville's story, although it may be an important background for interpreting it. Finally, however, Derrida is more interested in the linguistic properties and the grammatical effect of Bartleby's utterance, "I would prefer not to." Derrida notes that it seems like an incomplete sentence and he dwells on its inability to be completed.

Deleuze similarly insists on the linguistic strangeness of Bartleby's formulation. Deleuze remarks that it sounds like a "bad translation of a foreign language."[72] Sounding like a foreign language is, of course, not a bad thing for Deleuze. In fact, this collection of essays begins with an epigraph from Proust, who said, "great books are written in a kind of foreign language."[73] As does Derrida, Deleuze sees that Bartleby's "I would prefer not to" is "neither an affirmation nor a negation."[74] Bartleby is not refusing to do what he is asked, but he is not accepting the order either. There is a double negation involved, and because the only two possibilities are to say yes or no, Bartleby's impossible position of saying neither collapses into nothingness. Deleuze thus thinks that Melville goes Nietzsche one better. At the end of *The Genealogy of Morals* Nietzsche says famously that humans would rather will nothingness than not will. According to Deleuze, Bartleby is saying that he "would prefer nothing rather than something: not a will to nothingness, but the growth of a nothingness of the will."[75]

Could this nothingness be the basis of a politics? I do not see how. A politics implies a view of the future and a connection to the past. "Without past or future," Deleuze says of Bartleby, however, "he is instantaneous."[76] Moreover, Bartleby is a "pure outsider" who is *exclu*, and to whom "no social position can be attributed."[77] This should not be taken as a criticism, of course. Deleuze prefers to think of Bartleby as neither a particular case of a more general social trait, nor a type of literary character. "There is nothing particular or general about Bartleby: he is an Original."[78] On Deleuze's analysis, there can usually be only one such Original in each great work of literature. Originals, however, have no place in politics.

Derrida and Deleuze, on my reading of them, thus do not elevate Bartleby's utterance into an overall politics, and in particular, they do not attribute to Bartleby a politics of refusal. Slavoj Žižek, in contrast, particularly in his recent masterwork, *The Parallax View*, portrays and admires a political stance that he sees as Bartleby's gesture of refusal. Žižek has emerged as a leading critic of post-structuralism. Moreover, he sees himself as standing in but also going beyond the tradition of critical theory. Thus, when he makes a statement to the effect that symbolic fiction "structures our experience of reality," he is echoing critical theory, which has a long history of exposing social fictions that have had detrimental social effects.[79] Žižek likes to be provocative, and thus he argues in favor of a return to the Cartesian cogito in order to expose these social fictions. That is how he would correct the tradition of critical theory, which rejects the Cartesian cogito. But it turns out that the Žižekian cogito is not exactly Cartesian any longer since there is no discussion of mental substance. Furthermore, the cogito looks remarkably like the Lacanian unconscious. Žižek's conception of the self is closer to Kant's transcendental unity of apperception than to Cartesian mental substance, and in fact, Žižek defines the self as "this empty point of self-relating, not a subject bursting with a wealth of libidinal forces and fantasies."[80] In the metapolitical sphere he is strongly critical of what he refers to as the "oriental" reading of authenticity as a prescription for inner peace, no matter what is going on outside.[81] Accordingly, he also has objections to what he calls "western Buddhism," and its position that "social reality is an illusory game."[82]

If Žižek's stance is a Bartleby-like "Politics of Refusal," it does not seem all that different from "western Buddhism." There are, of course, some differences. Western Buddhism rejects all social reality and counsels complete withdrawal. After Heidegger's bitter experiences it is not surprising that he took a turn in the same direction. But if Derrida's reading, at least as I have characterized it, is right, Heidegger did not go as far as Žižek thinks in withdrawing from engagement. Despite my rendering of the later Heidegger in the previous section as lapsing into political silence, I do not agree with Slavoj Žižek's characterization of the Heideggerian politics of *Gelassenheit* as a politics of "Resignation."[83] Žižek defines *Gelassenheit* as "withdrawing from engagement."[84] Although Heidegger did say that only a "god" can save "us" now, his critique of modern technology and its strategy of enframing as well as his chiding of moderns for trying to control and dominate nature indicate that he still saw his later philosophy as capable of critique.

If Derrida is right, then I would add that *Gelassenheit* should not be interpreted as "withdrawal" so much as "letting be." "Withdrawal" is still too voluntaristic, as if we could really escape our social and historical situation. "Letting be" means not trying to control everything, but it is not simply an inner attitude. "Letting be" is something that has to be practiced over and over, and is thus still in active relation to the affairs of the world.

Presumably the same could be true of Žižek's Bartleby politics. Note that Žižek's Bartleby is a different character from Melville's. Whereas Melville's Bartleby takes no interest in anything, Žižek's Bartleby takes an interest in everything. If he did not, then there would be nothing to refuse. Žižek's Bartleby is not simply saying "no thanks." Žižek's Bartleby gives the distinct impression of being "passive aggressive." Žižek distinguishes, however, between "aggressive passivity," which is always actively working to make sure that nothing changes, and "passive aggressivity," which changes everything by doing nothing.

But then the question becomes how to distinguish between the do-nothing of quietism (political indifference) and Žižek's politics of refusal. There must be something more to the story. As I understand Žižek's Bartleby, his refusal may seem to be demurring from political activity, but actually the demurral is sharply critical of the social and political institutions that it is "refusing." Furthermore, what is going on inside the Žižekian figure is the opposite of Buddhist peace or Kierkegaardian inwardness. Rather than being "relaxed" about time, in both the ordinary and the Bergsonian senses, Žižek's Bartleby is seething inside. Žižek's irony about liberals and his disdain for poststructuralism suggest passionate commitments of a kind that Melville's Bartleby could not have displayed.

Žižek's criticisms of democracy reveal a subtle sense for the underside of democratic rhetoric. His strategy is like that of the smuggler who could not be caught because what he was smuggling was the wheelbarrow with which he left the factory everyday. To those who would justify the Iraq war, for instance, by pointing out that the world is better off without Saddam Hussein, Žižek responds, "Yes, the world is better off without Saddam—but it is not better off with the military occupation of Iraq, with the new rise of Islamic fundamentalism provoked by this occupation."[85] To those who claim that life is better in a democracy, Žižek similarly mentions but does not himself avow the usual criticisms of modern democracy. These are (1) that democracy is not truly democratic since a minority can shift votes dramatically, and (2) that political agents claiming to have insight into the "true nature of things" tend to want to impose this insight on everyone else. Instead, Žižek wants to emphasize that democracy itself makes possible such anti-democratic strategies. His criticism is not simply that democracy contradicts itself by harboring antidemocratic tendencies in itself.

Instead, he suggests further that democracy has suicidal tendencies and that democracy subverts itself. Egalitarianism, for instance, may be a matter of renouncing privilege so that no one else will be able to have it either. This Nietzschean observation is that egalitarianism is sustained not by a desire to be equal to others by sharing with them benefits that one has, but by envy of those who might have more than one has. In other words, just as Nietzsche uses the notion of *ressentiment* to argue that Christianity is based not on love but on hate, the suggestion here is that egalitarianism may be based not on sympathy, but on envy. Similarly, democratic society may depend on *evaluation* not so much because of any innate sense of fairness as because of *ressentiment* of difference. If human rights mean that all subjects have the same value, and are all self-identical without differential qualities that justify different treatments,[86] then everyone has to be tested, whether through standardized tests or extensive personal interviews, so that their potential can be identified and categorized without reference to any special "symbolic identities."[87]

In short, for Žižek the claim that democracy is the best form of society available represents a privileged view that suspends the rules of democracy whereby no such privileged perspective should dominate. To Richard Rorty's idea of "cultural politics," Žižek would probably point out that the notion of what is "cultural" has value only in contrast to what is "natural," and that what is "natural" is already at stake in "cultural politics." Insofar as Rorty himself would grant that the distinction between what is cultural and what is natural is a "political" issue, I take it that he and Žižek would not disagree about the principle of cultural politics. Where their "war of words" would take place is over the question of the value of democracy. Rorty does believe that democracy is the best form of society currently available. Žižek would probably refuse such a claim, not because he is opposed to democracy, but because he believes that its connection to global capitalism needs to be made clear. Like Bartleby, Žižek does not so much reject democracy as he refuses to accept it. Bartleby's "I would prefer not to," says Žižek, is not simply "the necessary first step which, as it were, clears the ground, opens up the place, for true activity, for an act that will actually change the coordinates of the constellation."[88] Žižek's point is not the simpler Hegelian claim of Hardt and Negri that Bartleby politics is the abstract negation that precedes more concrete determinate negation. Instead, Žižek's picture is more complex. For him, there is a continuous parallax shift between Bartleby's passive gesture of withdrawal and the active formation of a new order whereby the former "forever reverberates" in the latter. The refusal is "the very source and background of this order, its permanent foundation."[89] Both perspectives are required to see where we are to go, even if what is seen from one is different from the other viewpoint. Despite his dislike of the word "resistance," I conclude that Refusal is thus a central weapon in Žižek's repertoire of critical resistance.

Reflections

To sum up these phenomenological and postphenomenological analyses of the future, I note that although in principle this type of analysis could go into the different philosophers' sense of their own times and historical possibilities, that would be interesting primarily from a biographical point of view. Here I am interested in the question of whether there are necessary connections between the analyses of the temporal flow on the one hand, and social, political, and historical positionings, especially in our own times, on the other. In contrast to Heidegger, who prioritizes the future, many of the other theorists described in this book place more emphasis on the present. Or to be clearer, every philosopher who is concerned with the question of action will emphasize the present, because that is where the action is. Their attitudes toward the past and the future will depend, in turn, on whether the past and the future encourage or inhibit present action.

This return of the present may be the first sign that continental philosophy is moving out of the period of the "post." Terms such as postmodernism, poststructuralism, post-Marxism, and posthistorical make philosophy into a "late" or a "belated" social phenomenon. In contrast, emphasizing the present suggests that philosophy is moving into a different historical moment—one that it is still too soon to label definitively.

Of the theories of temporality discussed in this book, Bourdieu's might appear to put the most emphasis on the past. The bodily habitus incorporates dispositions that are then projected as expectations for the future. The habitus is thus a strongly conservative force. The habitus explains why we find certain patterns of action intelligible and why only specific actions seem plausible given the social field. To say that the habitus is strongly conservative is not to say, however, that Bourdieu's theory of the habitus is conservative. Although we are deeply entrenched in our habitus and thus in the past, Bourdieu thinks that reflective sociology can contribute to active social change by letting the appearance of social necessity become apparent as just that: appearance. Bourdieu depicts necessity so strongly because he knows that we resist the appearance of necessity and that once sociology reveals it, we will take action against it. The point is not simply to become reflectively self-conscious or self-critical. We "become who we are" in the present not so much by changing ourselves as by changing our world. Therefore, I read Bourdieu as criticizing Heidegger's account of the futural in order to prioritize the present.

By removing the vestiges of teleology, Derrida too can be read as emphasizing the present as the time of our lives. At the same time, he does not advocate the inauthentic present with its attitude of "wait and see" ("*voir venir*"). As he explains in *A Taste for the Secret*, the future erupts in the present unexpectedly: "the future rushes onto me, comes onto me, precisely where I don't even expect it, don't anticipate it, don't 'see it coming.' "[90] Derrida's analysis of temporality temporalizing itself should help to rebut criticisms of deconstruction for political quietism. One version of quietism is the reactive resistance of the sort that is labeled as "reformist." Reformists are accused of being averse to the prospects of revolution. Derrida insists, however, that just as deconstruction is not utopian, it is not also antirevolutionary, and it can invoke the rhetoric of revolution.

These remarks show either that it is a misconstrual of poststructuralists to think that they reject the rhetoric of social progress entirely, or that it is incorrect to label Derrida as a poststructuralist or a postmodern. Critics tend to think that anyone who is designated as "post" should reject the Marxian story of class struggle and revolution. In "Marx and Sons" Derrida says that he does not reject either the idea of class, however problematic it is, or the figure of revolution, however complicated he finds it. He insists that to label him as either a poststructuralist or a postmodern tends to oversimplify his theory of temporality and historicity.[91]

Even if one grants that point, however, there is a lingering issue with his account of messianicity. Abandoning teleology altogether threatens to make the messianic interruption into a moment of absurdity, where the totally unexpected erupts on the scene. Derrida's response to this threat of absurdity is to insist that the notion of democracy-to-come emphasizes not the distant future but the need to act here and now. From the practical standpoint, however, it is a legitimate question to ask whether there would have to be some more definite reasons from which to act, and collective goals toward which to aspire.

Like Derrida, Benjamin also emphasizes the present, and thus of the need for action, but with more normative bite. In contrast to Heidegger, Benjamin does not tie his account of the connectedness of life to the future or to death. The messianic moment can erupt at any point, but it is motivated more by past enslavement than by future liberation. The priority is on the possibility of action in the present, and as in the case of Derrida, there is a suspicion of putting off until the future the immediate need for social change.

How, then, should the phenomenology of futural temporality be understood? From the point of view of the metaphysics of time, the past and the future do not seem really to exist: the past because it is always over and done with, and the future because by definition it always has not occurred. In contrast, this chapter has tried to show that from the phenomenological point of view the past and the future in fact do exist, precisely as features of what could be called either the "lived present" or "the time of our lives." The past can be viewed differently, for instance, by reinterpreting the present. Moreover, the future is equally open to interpretation through action. The futural can be understood both as the projection of a present that is already past, and as the future of a past that has not occurred. As exemplified by Benjamin's angel, the future may not really be a function of what lies ahead of us. Instead, it might well be a function more of what lies behind us, as a possibility that once was to be realized, but that also exceeds what was once present.

5 *Le temps retrouvé*: Time Reconciled

There is the moment when a distinction is made and the moment when it is taken back. In these final pages I explore various strategies for reconciling lived temporality with objective time. This process involves rejoining the concepts of time and temporality, which I began by distinguishing. By the terms "reconciling" and "rejoining" I have in mind, of course, what Proust means when he speaks about "*le temps retrouvé*."[1] When he speaks of *le temps perdu* he does not mean only "past" time. Despite the French expression for wasting one's time, "*perdre son temps*," I do not think that he means "wasted time" either. As I read him, he is addressing what I designate as the "sting of time." This is the sense we have of being *in* time, of being subject to time's passing, and of being concerned with the fact that our lives are running out of time. In this sense, which perhaps anticipates the development in the 1940s of French existentialism, the "time of our lives" is an existential issue for each of us.

Of course, we should not forget the primary sense of the expression, "we are having the time of our lives." In ordinary parlance, this expression means that we are enjoying ourselves, that we are having a good time, perhaps the best time, of all the times we have ever had. Enjoyment is itself a temporal dimension, and, as the philosopher Levinas urges in *Totality and Infinity* (1961), the concept of enjoyment too should be included in any analysis of the connectedness of life. After all, having the time of our lives goes a long way toward making life worth living.

The task of reconciliation is to fuse these two senses of the "time of our lives." Reconciliation between the sting of time and the enjoyment of life has always been a goal of both literature and philosophy. Proust's *"temps retrouvé"* is a literary effort to reconcile us to the inevitability of time becoming lost and the power of reminiscence in retrieving it. "Reminiscence" is possible on a Bergsonian premise that the past coexists with the present. But Proust represents only one way that reconciliation can be envisioned. There are also other attempts at reconciliation by philosophers as far from one another as, for instance, Heidegger and Bergson, who try to ground time in temporality through quasi-transcendental arguments. These arguments depend on a distinction between primordial and derived. I also consider different philosophers' accounts of reconciliation that do not depend on transcendental arguments, logical priority, or a priori status. These include Nietzsche, Deleuze, and Žižek, among others. Grouping the theories according to historical affinities gives us the following four debating arenas.

First, there are those who seek reconciliation through memory. Proust with his notion of "reminiscence" is a paradigm case of this approach. Deleuze characterizes Proustian reminiscence as involuntary synthesis. In contrast to this involuntary, passive synthesis stands Walter Benjamin's voluntary active synthesis of "remembrance." These two different kinds of memory rely on different accounts of the relation of time and temporality, as I will show below.

The second group consists of Husserl and then Heidegger in his different stages. Heidegger adapts Husserl's structure of "retention" into "resoluteness" in *Being and Time.* Alienation from time is manifested in inauthentic Dasein by the mechanics of what I call "regulation." The later Heidegger has a more passive relation to time, which he calls *Gelassenheit,* and which Žižek interprets as withdrawal from engagement, or "resignation."

The third grouping represents recent attempts to move beyond critical theory, particularly as practiced by the Frankfurt School and its successors, including Jürgen Habermas. One goal of this group is to reconcile past and present by writing what Foucault calls the "history of the present" (that is, the critical history of how we have become who we are) without appeal to the notion of ideology, at least in its classic sense as false consciousness.[2] The epistemological problem with ideology in this sense is how a class consciousness can be said to be false if it is unclear how a class consciousness can be said to be true. Slavoj Žižek wants to rethink the idea of ideology without abandoning the possibility of critique. More radically, Michel Foucault and Gilles Deleuze have little or no use for the idea of false consciousness, and they believe that social criticism is nevertheless still possible without appealing to the notion of ideology at all.

The fourth grouping shows how Husserl and Bergson, without being merged, nevertheless can be reconciled in Deleuze's account of Aion and Chronos. Nietzsche and Bergson are also brought into dialogue. Nietzsche seeks reconciliation through "recurrence." Bergson assimilates past, present, and future in the process that he calls "relaxation" (as opposed to contraction). For Deleuze these strategies all represent what he identifies as "repetition."

Not every theorist mentioned in this book is listed here, and no definitive list of strategies is possible. The point is to recover several different ways of merging time and temporality so that the possibility of reconciliation of temporality and time is at least indicated. Each of these strategies has its advantages and disadvantages, and none can claim to be more than a possible interpretation of the problem. Each is thus a possible solution or resolution, although not the only one. Each one may see itself as the synthesis of the others, but the sequence of the groups is not apocalyptic and there is no prioritization implied. No one theorist gets it right and no one is wrong. These are different ways of dealing with different senses of the time of our lives. They are all possible strategies, even if ultimately each will fall short. The sting of time can never be taken away entirely. If we can reconcile ourselves to that fact, then we will have made a positive step toward living more completely.

There is also a methodological side to this reconciliation. Phenomenology is not a subdivision of the philosophy of mind. If phenomenology is going to be able to rethink philosophy in a thoroughgoing way, it should not be thought of as analyzing "mental" experience only. The world is as temporal as the mind and temporality is not to be thought of as coming exclusively from one or the other. That is to say, temporality should not be put in the box of either idealism or realism.

What we are witnessing throughout this discussion of the reconciliation of the sting of time with the enjoyment of life is also the reconciliation of the difference between descriptive phenomenology and normative hermeneutics. No longer merely descriptive, phenomenology turns into genealogy insofar as it does not shirk the normative issues at stake. These normative issues can be existential, moral, social, or political, but they need to be addressed. Genealogy, unlike phenomenology, cannot settle for explaining how *knowledge* is possible. If, as Levinas also charged, traditionally phenomenology is preoccupied with the cognitive, now its critical potential for explaining how normative comportment is possible must be developed.

One dimension of the normative is the political. By the "political," I mean a philosophical account of how normative discussion in the social realm or the "public sphere" can occur.[3] Politically, the various theories of how temporality temporalizes itself, whether from the future or from the past, have had consequences for an understanding of agency and action. Although there is little endorsement here of the hope built into Kant's optimism about the progress of humanity toward the best possible society, Derrida's insistence on his right to irony is a good example of how activism can be justified even by a more pluralistic and less universalist theory. Irony can, of course, lead to a do-nothing cynicism. But a more thoroughgoing irony can also turn cheerfully against itself. Thus, the irony of the present study is that although it calls for engagement in the social, political, and historical world, it does so at such a basic philosophical level that its relevance for political activism may be obscure. For analysis of this level of generality the only justification is that it makes action more intelligible. Unlike an event, which is a mere occurrence, an action requires some understanding of what it intends to do or to achieve. Activism requires action, but to count as such, an action must be understood. In these studies I am exploring philosophical views about the most basic condition for the possibility of action, political or otherwise, namely, temporality. Action presupposes time, so any theory of action must contain, at least implicitly, a view about time.

Applying the method of genealogy to temporality sweeps away some of what common sense believes about time, and at the same time it recoups and preserves certain insights that are buried in our ordinary and mythical ways of talking about time. As Merleau-Ponty remarks, "There is more truth in mythical personifications of time than in the notion of time considered, in the scientific manner, as a variable of nature in itself, or, in the Kantian manner, as a form ideally separable from its matter."[4] Temporality is often invisible. But if these investigations awaken intuitions and quicken reflections about a phenomenon that is usually below the threshold of visibility, they will have served their purpose.

For example, one way to test any account of temporality is to see its critical potential over against some standard platitudes of how to deal with the sting of time. In comparison to such adages, the much more complex strategies of reconciliation that I describe below will show up to an advantage. Take, for instance, the injunction to "live for the moment." The idea is to live without any regard for or thought of the future. If Husserl, Heidegger, and Bergson/ Deleuze are right, this would be impossible advice to heed. One must always have some future. Indeed, this advice to live as if there were no tomorrow is really saying that the near future is simply more important than the more distant future.

Arthur Schopenhauer, philosophy's wittiest pessimist, dispatches this adage handily. To the allegedly greatest wisdom that would make the enjoyment of the present, which is the foundation for the entirety of our existence, the supreme object of life and thus the only reality, Schopenhauer retorts that this motto is really the greatest folly. "For that which in the next moment exists no more, and vanishes utterly, like a dream," he says, "can never be worth a serious effort."[5] Similarly, once something is in the past, for Schopenhauer, it loses all reality: "That which *has been* exists no more; it exists as little as that which has *never* been."[6] Schopenhauer's attitude toward temporality is captured pithily when he writes, "Time is that in which all things pass away; it is merely the form under which the will to live . . . finds out that its efforts are in vain; it is that agent by which at every moment all things in our hands become as nothing, and lose any real value they possess."[7]

Pessimism indeed can be an effective strategy for dealing with the sting of time. By recognizing the problem as the truth of the matter, and by adopting an ironic stance toward this supposed truth, pessimism puts itself forward as the most consistent stance possible. There is, however, a difference between pessimism and pathology. There are cases of pathology induced in people who are forced to live just for the moment if only because it does seem unlikely that there will be a tomorrow. In the concentration camps of the Holocaust, for instance, or in the grips of severe joblessness or hopeless social circumstances, people can live for the moment without any hope for the future. This social pathology induces psychological pathology that must appear to itself to be normal insofar as there are no other viable solutions in such circumstances.

The alternative adage to "live for the moment" is always to "plan for the future." Much loved by parents and hated by their offspring, this saying has the obvious disadvantage of sacrificing present satisfactions for abstract future securities and benefits that may never materialize. Is there anything comparable about the past? Proust and Benjamin point to the ameliorative power of memory. Even memory can become exaggerated and painful, however. In Jane Birkin's recent film, *Boxes,* her character cries out, "I can't take any more memories!" In the political sphere, the emphasis on memory and tradition can become nostalgic to the point where action is only ever undertaken to maintain the status quo. These platitudes, "live for the moment" and "plan for the future," show how ignoring the dimensionality of temporality can lead to questionable experiential generalizations.

This chapter is, then, an attempt to provide an example of how the phenomenological analyses could be applied to existential, normative issues. To help in keeping track of the complex terrain of the discussion, an outline is provided in figure 5.1, with the proviso that the groupings are for convenience of exposition only and that other combinations are appropriate as well.

Strategy 1   Remembering: Proust and Benjamin

In the course of explaining Bergson in *Difference and Repetition* as well as in *Proust and Signs,* Deleuze raises the question, how can we reconcile ourselves to all the time that we have lost and the little bit of time that is left to us? How can we redeem the past? This dilemma is familiar from Proust's major work, and so is Proust's answer: through "reminiscence." Proust had studied with Bergson, so it is appropriate for Deleuze to translate this notion into Bergson's terminology as a passive synthesis, that is, an involuntary memory that differs (in kind, not simply in degree) from the active synthesis of a voluntary memory. Usually Proustian immanence is explained as the association of a past and a present sensa-

---

Reconciliations of Time and Temporality

I. Recovering lost time through memory
Reminiscence—Proust
Remembrance—Benjamin

II. Recovering lost time through interpretation
Resoluteness—early Heidegger (authentic)
Regulation—early Heidegger (inauthentic)
Resignation—later Heidegger
Retention—Husserl and Merleau-Ponty

III. Recovering lost time through critique
Revolution—Marx
Reflection—Foucault
Rogues—Derrida
Refusal—Žižek

IV. Recovering lost time through temporalization
Recurrence—Nietzsche
Relaxation—Bergson
Repetition—Deleuze

Figure 5.1

tion. But this theory of the association of ideas cannot account, Deleuze believes, for the intense joy that Proust indicates. Combray, the childhood village of the protagonist, comes back to Proust's protagonist suddenly from the taste of the tea and the madeleine. In *Difference and Repetition* (and similarly in *Proust and Signs*) Deleuze describes this moment of reminiscence poignantly as follows:

Combray reappears, not as it was or as it could be, but in a splendor which was never lived, like a pure past which finally reveals its double irreducibility to the two presents which it telescopes together: the present that it was, but also the present present which it could be. Former presents may be represented beyond forgetting by active synthesis, in so far as forgetting is empirically overcome. Here, however, it is *within* Forgetting, as though immemorial, that Combray reappears in the form of *a past that was never present*: the in-itself of Combray. . . . The present exists, but the past alone insists and provides the element in which the present passes and successive presents are telescoped.[8]

The village is remembered not as it was, but now in its significance for all his life, which he could not have realized as a child. In that sense, then, Deleuze says that this past has never been present.

Spontaneous reminiscence is not something that one can willfully attain. As a form of coming to terms with the passage of objective time, a Proustian reminiscence is not something that one can simply decide to have. Moreover, reminiscence is a radically individual affair that will be different for everyone. Many of us have had such an experience at least faintly, perhaps by revisiting places once familiar in one's youth. Such experiences require some temporal distance from the past, but not necessarily a great deal. Going back to Paris year after year can, for instance, involve reminiscence of previous visits—in a form of what I referred to earlier as a *wirkungsgeschichtliches Bewusstsein*.

In *Proust and Signs* Deleuze presses the question of where the intense feeling of joy comes from, "a joy so powerful that it suffices to make us indifferent to death."[9] The mere similarity of the taste of the madeleine on two occasions is not a sufficient explanation for the strength of the emotion that Proust describes. Deleuze begins to answer this question by noticing that *voluntary* memory is standardly theorized as connecting an actual present with an actual past. But for any Bergsonian, this way of thinking about memory misconstrues the phenomenology because it first breaks temporality up into moments that are like different photographs that are then compared. Proust rejects this photographic metaphor for the reason that it misrepresents the essence of pastness. The metaphor of the past as like a series of photographs, or a succession of cinematographic frames, loses the sense in which time *passes*.

The past does not get frozen into a frame the instant it is over. For Bergsonians like Proust and Deleuze even Husserl's account fails to capture the sense in which the past is not a real succession of instants, no matter how interconnected. Instead of a real succession (*une succession réelle*), the Bergsonian past is constituted, on Deleuze's reading, in a virtual coexistence with the present (*une coexistence virtuelle*).[10] The "virtual" is precisely the essence of pastness, which is not to be a past that "has been," but to be something that "is" and that coexists with the present. If voluntary memory breaks the past and the present into separate domains, involuntary memory shows their more primordial immanence, their inherence in each other at least virtually if not in reality, or in their "truth" if not in their actuality.[11]

There are two phenomenological features of involuntary memory that should be emphasized: its abruptness and its brevity. First, involuntary memories can break into the present in sudden, unexpected ways that could not be prepared for or anticipated.[12] The "pure" past cannot be reduced to any single present, and it exceeds each and every present. Second, involuntary memories surge up, but do not last long. Indeed, they could not last long. Like a glimpse of eternity, they cannot be more than instantaneous. Nevertheless, the effect of the experience can be so valuable that it is said to make life worth living. The brevity of the moment of the madeleine contrasts markedly, then, with the length of Proust's novel. Like Bergson's cone, the taste of the madeleine contains in a "contracted" manner the entire novel, which is achieved with the gradual unpacking or "relaxing" of the singular experience.

The lost past recovered through reminiscence carries with it a sense of the entire life in which it figured. The length of the novel is due to the fact that the entire life must be experienced to realize the significance that Combray had for this particular life. The significance of the memory for the entire life is not something that could be remembered as simple recall or "recollection." The memory's connectedness to the present surges up along with the past scene.

Reminiscence is, of course, a form of redemption that is *aesthetic.* The aesthetic feature does not mean, however, that involuntary memories are *fictional.* If reminiscence can change our lives by giving us a different sense of the story of our lives, that story is not a fairy tale. Whether or not spontaneous reminiscence results through a form of sensory stimulation or synesthesia, as it did for Proust through the madeleine and tea,[13] it is also *personal.* As aesthetic and personal it might seem to be open to the objection that it has little to do with the *political.* In fact, when reminiscence is made into a philosophical category, it runs the risk of losing the redemptive value it had as a literary experience. Stated as a philosophical strategy for personal redemption from the passing of time, it seems to presuppose alienation and anomie, and of course it could appear to be nostalgic.

In chapter 3 the question was raised, is nostalgia always bad? The standard objection to nostalgia is that the object of nostalgia is a past that never was. This past is usually highly personal and accessed through private reminiscence. As a purely personal form of reconciliation, reminiscence has other limitations as a strategy for reconciliation than its lack of relevance for or interest to others. Memories that surge up may also be painful and of limited value for reconciliation. Repressed memories of childhood abuse or other forms of trauma are more often the cause than the solution to *ressentiment* against time. Loss, as of parents, can also be painful even late in life.

A response to these objections is apparent in the examples at hand, namely, Proust writing his quasi-autobiographical novel, and Bergson, with his notion of the past coming into being as its present. Bergson is certainly not nostalgic, and neither is Deleuze, affirms Alain Badiou in his frequently critical book, *Deleuze:* "*La clameur de l'Être.*" To be nostalgic is to have a sense of a loss of being, and a negation of becoming. In contrast, the temporality of

Bergson and Deleuze involves a sense of increased, supplemented being ("*un accroissement, un supplément d'être*").[14] Temporality is a "double creation," a "*scission créatrice*," that generates not only the past but also the future.[15]

Proust is hardly nostalgic in the above sense, either. For one thing, the pain of the past is not covered over. Furthermore, writing is itself a way of coming to terms with time, and Proust's writing in particular is a way of confronting one's present and future through memorialization based on one's own sense of the past. Proust is unique, of course, for the delicacy of his reminiscences. Moreover, recreating his life as an aesthetic experience has had social value and is not in fact an act of isolated individualism. Nietzsche is another example of someone who reworked his life into his writings and thereby reconciled himself to time.[16] Writing, whether literary or philosophical, is a way of coming to terms with time. There is no guarantee of success, but then, no one succeeds in overcoming time in the end. All one can hope for through this individual effort is reconciliation with one's own temporality, the time of one's own life.

A contrasting attitude is "remembrance." This notion is derived from Walter Benjamin. According to the account of memory of Bergson as transmitted through Deleuze, memory is an active and voluntary effort to recapture a past that is already at a distance from the present moment. Remembrance is thus different from reminiscence on a past that was never present. Benjamin's strategy is a voluntary way of achieving an effect similar to Proust's, but on a politically broader base. Benjamin thinks that social and political resistance is motivated more by past injustice than by future hope. There is thus a weak messianic power involved in remembrance insofar as it reminds us of past expectations that have been lost or buried by other interpretations of the tradition. Although remembrance can be personal and individual, it can also be social and collective. We all remember individuals who have been important not only to ourselves, but to others. Friends, teachers, idealized others are not just examples, but paradigms—model existences. Benjamin's life is itself a paradigmatic example of how remembered injustice could generate present action to bring about a better future for others and not merely for oneself. By writing history from the point of view of the victims—against the grain, as it were—time is not reversed, but its course is deflected. Remembrance can remind present agents that there were forks in the road with paths that were not taken, but that are still open today through revisionary thinking. The power of tradition is not identical to traditional power.

Benjamin's weak utopian moment presents a challenge to leftist philosophy of history insofar as it calls for the reconciliation not only of time and temporality, but also of the temporal and the historical. The Italian philosopher Giorgio Agamben maintains that every conception of history carries with it a specific conception of the temporal. What Benjamin begins and Agamben continues is the elucidation of this implicit sense of temporality. Agamben's belief is that no new society is possible without a reconceptualization of time. Modern political thought has been so preoccupied with the philosophy of history, Agamben maintains, that it has neglected the concept of time. This neglect erodes its ability to visualize the distinctive political possibilities of the modern era. "The original task of a genuine revolution," Agamben argues, "is never merely to 'change the world,' but also—and above all—to 'change time.' "[17]

Agamben sees Benjamin and Heidegger as the thinkers who have articulated the modern sense of time in contrast to the ancient and early Christian conceptions of time. Whereas the Greeks had a circular conception of time and therefore had little sense of history, the early Christians are said to have had a linear view of time that allowed for the development of a sense of historical direction. For Agamben, the modern conception of time is a secularization of Christian linearity. Both Benjamin and Heidegger base their critiques of linear time on a displacement of "continuous, quantified time."[18] Agamben sees these philosophers as opening the door to an understanding of time that is revolutionary, not only in the sense that it is novel and unanticipated, but also in the sense that it is the conceptual model for revolution: "It is this time which is experienced in authentic revolutions," writes Agamben, "which, as Benjamin remembers, have always been lived as a halting of time and an interruption of chronology."[19] For Agamben, Benjamin's "Theses on the Philosophy of History" is especially significant in that it shows that "history is not, as the dominant ideology would have it, man's servitude to continuous linear time, but man's liberation from it: the time of history and the *cairós* in which man, by his initiative, grasps favorable opportunity and chooses his own freedom in the moment."[20] Insofar as Agamben singles out Heidegger as the other discoverer of the modern sense of temporality, I now turn to Heidegger's understanding of how to regain lost time.

Strategy 2   Interpretation: Heidegger and Hermeneutics

Philosophical hermeneutics also analyzes the tradition as providing critical power in resisting present interpretations of who we are and how we have become this way. Some dialogue will thus be possible between Benjamin and Proust on the one hand and Heidegger and Gadamer on the other. Of course, the political differences of Heidegger and Benjamin in their personal lives will also be relevant to this conversation. Heidegger's rectorship and his affiliation with National Socialism cannot be easily separated from his philosophy. Not only his personal life but also his philosophy reflects his involvement on the wrong side of practical politics starting in the early 1930s. Before that he seems to have been mainly caught up in his work. Agamben maintains, however, that as far as Heidegger's notion of primordial temporalization goes, this conception of the temporal is "in no way opposed" to the Marxist philosophy of history.[21] If that is right, then Heidegger's phenomenology of temporality can in fact contribute to the analysis of the modern *Zeitgeist*. This analysis becomes richer and deeper, however, when Bergson's views are incorporated into it.

Heidegger takes over Husserl's notion of retention as a way of understanding how the past accompanies the present as new presents push old ones into the background. We have seen in chapter 3 the limitations of Husserl's account, especially when it is contrasted with Bergson's account. Heidegger's deliberate lack of engagement with Bergson on the topic of temporality is disappointing, then, because Heidegger's notion of the ecstases of temporality seems in many ways more like Bergson's cone than Husserl's linear diagram. Ecstases are not coming out of the past or the future, but are fully present. They indicate in which direction temporality is oriented. Note that there is not one direction for all of time, but each ecstasis has its own directionality. The ecstasis can move outward toward the past by "relaxing" the focus on the need for present action. Or it can enhance present action by focusing on the near future in a more "contracted" way.

Where Heidegger differs most significantly from Bergson is in (1) the priority that Heidegger gives to the future in contrast with Bergson, for whom the future is not real, and (2) the account that Heidegger gives of historicity and temporality. As I have argued earlier, Heidegger's emphasis on the future is often wrongly attributed to his fascination with death. What really motivates his privileging of the futural is not death, but his hermeneutical account of understanding and interpretation. Insofar as the understanding is always forward-looking, it involves what Heidegger calls the "projection" of possibilities. Unlike Husserl and Bergson, who have relatively little to say about the future, Heidegger's orientation is largely futural. The overall tenor of Heidegger's forward-looking philosophy thus contrasts with Bergson's and Husserl's more past-oriented theories.

As for the second difference between Heidegger and Bergson, interpreters of Heidegger often think of Dasein as temporal most primordially, and then with historicity added on as a secondary, optional layer. On my reading, this idea of layers misconstrues the relation of historicity and temporality. The assumption that I wish to counter is that Dasein is most primordially a singular conscious individual who exists autonomously from others. I take Heidegger at his word that Dasein is always already Mitsein, or being-with-others. Insofar as the Mitsein involves historicity, what leads Heidegger to describe temporality first and historicity second is the order of explication. He first lays out the modes of temporality required for life to be connected from birth to death. But historicity is equally required by the order of being insofar as Dasein is part of a social world right from the start. So although he first explains temporality and then historicity, in fact Dasein could not be a temporal being unless it were also capable of being a social being. Earlier I remarked that temporality is not necessarily historical because it seems possible that people might be aware of time passing without any awareness that history was in the making. That remark does not prevent me from asserting now that any temporal being is also *capable of* political engagement and historical involvement. The limitations on the possibilities for concrete action are of course part of the situation. Once again, however, remembrance shows us individuals who managed to act historically in socially significant ways even against overwhelming odds.

Mention of historicity raises the specter of relativism. For instance, Heidegger's predecessor, Wilhelm Dilthey, often relativizes knowledge and values to worldviews. Heidegger's insistence on the historicity of Dasein does not lead to the relativistic historicism that Dilthey encounters with his account of worldviews. The worry is that any theory involving the idea of different worldviews has trouble explaining why some worldviews are better than others. If value is relative to worldviews, then what it is wrong for me to do might very well be acceptable to someone from another culture. Note, however, that the idea of a worldview depends on the *scheme–content* distinction. The Kantian tradition took Kant's idea of the twelve categories of experience to imply that if different concepts were involved, a different conceptual scheme would produce different experiences, and indeed, even a different world from our world. There would thus be a certain incommensurability between conceptual schemes. Talk of conceptual schemes is thus relativistic. It is important to realize that Heidegger's appeal to the notion of interpretation *breaks* with the neo-Kantian faculty psychology whereby alternative conceptual schemes can be imposed on a given content. Unlike conceptual schemes, interpretations are not incommensurable. Heidegger does not need to think that his interpretation of human existence is relativistic. He can consistently believe that he is interpreting human existence, and that his interpretation is sound.

Historicity is a permanent feature of human existence, just as are the structures of understanding, attunement, and discourse. Heidegger can believe this without contradiction because although the Dasein is always historical from an ontological or existential point of view, it is not always historical from an everyday (ontic or existentielle) point of view.[22] That is because Dasein's historicity can be undifferentiated, authentic, or inauthentic. To be undifferentiated is not yet to be either authentic or inauthentic. Not all peoples have recognized their historicity. If Hegel were right, a people that did not have writing would not be able to interpret itself as historical. (Heidegger refers to such people as "*primitive* Dasein"—"primitive" is, however, a term that is challenged today on the grounds that it simplifies people who are just as complex as moderns, if not more so.[23]) These people are not inauthentic, however, but merely undifferentiated between the authentic and the inauthentic. Inauthentic historicity results from a narrowing down of the range of possibilities to a few that are simply taken over from the past. In contrast, authentic historicity is achieved through critical awareness of the wider range of possibilities that are open at a given moment. This critical awareness applies to one's own situation as well, and Heidegger says that one must always be willing to "take back" (*zurück-nehmen*) one's resolve to pursue certain possibilities if that resolve leads one into a seriously compromised position.[24] Of course, Heidegger should have remembered his own notion of a self-critical take-back later in the 1930s when he apparently regretted, although he never renounced, his decision to ally himself with National Socialism.

Heidegger goes beyond Husserl's notion of retention insofar as Heidegger is not concerned with the linear, diachronic connections between discrete moments. Husserl's diagram commits Husserl to try to connect the dots of the series of nows to make a straight line. That is, Husserl's spatialization of time into distinct nows is built into his sense of what his philosophical task is and what he has to show. In contrast, Heidegger's ecstases are built into the present in such a way that time and temporality come together again, if only to come apart right away in the distinction between inauthentic and authentic temporality. Inauthentic temporality reduces time to a series of nows through the process of "leveling off," or what I call "regulation." Heidegger says that inauthentic temporality must think of time in terms of nows that are diachronically the same, and quantitatively identical. Authentic temporality, in contrast, does not reduce time to measurable instants, but sees time and temporality as conjoined in a qualitative and not merely quantitative sense of time's duration.

Against the measurable, quantified time that regulates and controls inauthentic Dasein, Heidegger contrasts the temporal duration built into the moment of vision of authentic Dasein. At this point, however, it becomes important to distinguish the early from the later Heidegger, that is, the Heidegger of the middle to late 1920s, including *Being and Time* in 1927 on the one hand, and the Heidegger of the "Letter on Humanism" of 1947 as well as much of his writing after that. The early Heidegger argues that authentic Dasein can take over its finitude explicitly through what he calls "anticipatory" resoluteness. In anticipatory resoluteness, authentic

Dasein does not simply sit back and *await* the future, but actively *anticipates* it by going out and doing something about bringing it about. This response indicates that authentic Dasein has to *answer for* the world actively.

The contrast between the early and the later Heidegger is brought out nicely in the later Heidegger's discussion of the sentence from *Being and Time*, "*Es gibt Sein*" (literally, "it gives Being," although it is better translated as "There is Being"). The early Heidegger was often read as meaning by the "*Es*" that the giver of Being was *Dasein*. In 1947, however, the later Heidegger maintains in the "Letter on Humanism" that when *Being and Time* says "*Es gibt Sein*," the "*Es*" is *Sein*. In the early text, the sentence in question reads, "Only so long as Dasein is, is there [*gibt es*] Being." The later Heidegger claims that because Being is both the clearing and what sends the clearing, Dasein was not meant to be the clearing:

For the "it" that here "gives" is Being itself. The "gives" names the essence of Being that is giving, granting its truth. The self-giving into the open, along with the open region itself, is Being itself.[25]

For the later Heidegger, Being and not Dasein is thus the basis for the open clearing in which things show up.[26]

Although these might seem to be metaphysical issues that have little practical bearing, the later Heidegger gives up the activist stance he takes in the 1920s, before his political disillusionment. In contrast, he adopts what I call a stance of ontological passivism. The stance is ontological because it involves the clearing itself as the condition for what shows up in the clearing. The stance is passive because on this later view, Dasein *receives* its clearing from *Sein* instead of actively taking on the world of its own accord. Criticism is possible only through poetry and philosophy done more as "meditative thinking" than as social and political engagement. In fact, Heidegger warns against getting involved in the public sphere: "if humanity is to find its way once again into the nearness of Being, it must first learn to exist in the nameless. In the same way, human beings must recognize the seductions of the public realm as well as the impotence of the private."[27] There is little that we can actually do, the message is, and the best we can accomplish is simply to wait, to be ready, and to be open to new possibilities, should they materialize.

This passivism is reinforced by the later Heidegger's statements to the effect that humanity is merely the "shepherd of Being."[28] In contrast to Sartre's existential humanism whereby, in Sartre's words, "[humanity] is the being whose appearance brings the world into existence," Heidegger writes, "[humanity] does not decide whether and how beings appear, whether and how God and the gods or history and nature come forward into the clearing of Being, come to presence and depart."[29]

Which of Heidegger's accounts is preferable, the earlier one or the later one? Certainly there are advantages to each. His earlier stance of normative activism provides a more positive account of agency. Although limited by finitude, Dasein can take charge of itself and transform its situation to a certain degree. The attraction of this early account is that it is more optimistic about the possibility of the transformation of self and society. From the standpoint of the passivism of the later Heidegger, however, this treatment of agency looks like a continuation of the anthropocentric tradition that allows us to dominate and destroy nature. Rather than seeing "humanity" as the measure of things, the later view challenges the very idea of humanity in an even more radical way than the earlier view does.

Heideggerian hermeneutics is an area where time and temporality are reconciled in tradition. Connection to tradition makes the efforts of social life intelligible. Rather than being nostalgic, interpretation opens up avenues for future action that carries out values and choices that have been forgotten. Not reactionary, hermeneutics opens up possibilities. For some, however, a more radical political temporality will be desirable, and that is why the postcritical theory movement, to be described next, represents a more activist alternative.

Strategy 3 Critique: Foucault, Derrida, and Žižek

This section is concerned with political temporality, that is, with the way temporality conditions politics. From Marx, for whom revolution is the way that time and temporality are brought together, through the poststructuralists, who lose hope in the teleology of revolution, the radical political scene appears to many to lose its motivation and momentum. Žižek's Bartleby politics, discussed in chapter 4, may be the latest and most provocative attempt to revitalize an "in your face" style of politically activist theory. Whether it is any less quietist than the philosophical movements that it would replace, however, is still a question. After a discussion of Foucault's account of the temporalization of modern society, this section returns to Žižek's critique of both Heidegger and poststructuralism, because Žižek sees poststructuralism as being obsessed with the problem of Heidegger's politics.

First, I must explain how I distinguish poststructuralism from the period following poststructuralism in which we now stand. I date the beginning of poststructuralism from the publication of Deleuze's book on Nietzsche in 1962 and its end with Foucault's death in 1984. Even Foucault had moved beyond poststructuralism at that point. Poststructuralism shifts philosophical attention from Hegel, Husserl, and Heidegger to Freud, Marx, and especially Nietzsche. The structuralism that precedes poststructuralism is not a philosophical movement, but more of a social scientific methodology to replace philosophy with the so-called sciences of "man." Preoccupied with bringing these sciences up to the level of the sciences of nature, structuralism attempts to find rules for conduct that operate independently of and prior to conscious awareness and control. Poststructuralism continues this activity of giving explanations of controlling but invisible features of behavior.

Poststructuralism gives up, however, structuralism's pretensions to scientific objectivity as well as to philosophical foundationalism (that is, the attempt to base philosophy on indubitable premises). Although language is the paradigm for both structuralists and post-structuralists, poststructuralist philosophers supposedly no longer place as much emphasis on universal rules, whether a priori or empirical. Instead, they tend to theorize difference rather than identity, multiplicity rather than univocity, simulacra rather than similarity. A list of the main doctrines of poststructuralism would have to include the following: (1) critique of subjectivity; (2) critique of universality; (3) critique of progress in knowledge; (4) critique of progress in history.

Michel Foucault

Poststructuralists tend to avoid normative philosophy and they are wary about normative standards. In the 1980s, however, Foucault and also Derrida began to write on ethical, social, and political issues in positive, constructive ways. Perhaps their motivation for doing so came from the fact that they were frequently charged with being unable to discuss normative issues without contradicting themselves. They were often challenged to say in the name of what they were criticizing social formations. To show that he could in fact take on value-theoretical issues, Foucault moved first from the method of archaeology to the method of genealogy. Both of these approaches point to subliminal factors that make us who we are. Archaeology tries to unearth the structures or rules that govern our discourses. Genealogy uncovers the power relations that form our very subjectivity. Foucault then moves from writing the genealogy of power to writing the genealogy of ethics. In the interview entitled "On the Genealogy of Ethics," he adapts Hegel's notion of ethical substance to his studies of the history of sexual ethics.[30] He moves beyond thinking of social practices as being acquired below the threshold of perceptibility, and calls explicitly for the value of "reflection" on what we can do freely. Reflection might well show us that we are not as free as we think, but nevertheless without reflection there is no freedom.

From Gadamer we learn that what we are after is better self-understanding, and that who we are is a function of who we have been. From Benjamin we learn to think more radically of the past as a means of disrupting our sense of who we have been. From Foucault we learn more dramatically to unlearn who we have come to be. The Foucault-Benjamin line is summed up aptly by Elizabeth Grosz who writes, "The past is our resource for overcoming the present, for bringing about a future."[31] The past is the source not simply of the repetition in the present of what has been successful for those who dominated the past. The tradition also includes over-looked possibilities that, when seized, can disrupt present oppression and lead to a different future. We have to look more to lost, forgotten, or disguised traditions to find alternative futures for ourselves.

Foucault radicalizes tradition by way of his distinction between practices of freedom, which he favors, and practices of liberation, of which he is wary. He is suspicious of Wilhelm Reich's idea of liberation, which he thinks is derived from a dubious interpretation of Freud.[32] "Liberation" in this specific sense seems to imply that there is something already there, one's true self or one's innate sexuality, for instance, that has been repressed by power and that merely needs to be released. He also thinks that Reich's account of liberation rests on the assumption that power is only ever domination.

Practices of freedom, in contrast, do not entail this theory that all power is domination. All domination is power, but not all power is domination. Moreover, these practices of freedom involve the ability and desire to resist the micropowers that pressure us to conform. Foucault is not necessarily opposed, however, to *non*-Reichian senses of liberation. Despite his suspicions about the term, he will allow us to speak of liberation as long as these Reichian associations are avoided.

Similarly, when he speaks of freedom, he does not mean freedom as opposed to power: "I do not think," he writes, "that a society can exist without power relations."[33] But because power relations are not opposed to freedom, and because power is not necessarily domination, the point is to "play these games of power with as little domination as possible."[34] Genealogy may contribute to the process of breaking with our traditional sense of the past and who we are through what Foucault calls *désubjectivisation* or *désassujettisse-ment*, which could be translated as "desubjectification," "desubjec-tion," or "desubjugation." Foucault sees the past as working itself out in the present through the materiality of power. The materiality of disciplinary power is embodiment, and the materiality of bio-power is population. Genealogy breaks with the materiality of the past by resisting the identities that limit our possibilities. These identities make us who we are, but they are not fixed and inaltera-ble. If in the past we let them be imposed on ourselves, we can in turn alter them through transformative practices. Desubjectification through practices of the self is difficult but not impossible.

When Foucault talks about taking care of the self, he does not limit this process to individual subject identities. In fact, he challenges the idea that individuals exist as such before their socialization.[35] What he says about the self would thus apply equally well to collective, communal, or *social* subject identities. Desubjectification would thereby radicalize the traditions that form our identity. Critical resistance that promotes this kind of social desubjectification will open the door to social change resulting from the accompanying denormalization, depsychologization, and deindividualization.[36]

In response, then, to the charge that the theory of power cannot specify that in the name of which oppression is to be resisted, Foucault's message strikes me as being quite clear. If genealogy must specify that in the name of which resistance is justified, let it be in the name of transformative freedom, properly understood. After all, Foucault himself says that there are cases and situations where "liberation and the struggle for liberation are indispensable for the practice of freedom."[37] The desubjectification will not be followed by anarchy or barbarism, but by the formation of an ethics of how to live. Foucault defines ethics as "the reflected [*réfléchie*] practice of freedom," and he maintains that "freedom is the ontological condition of ethics."[38] In short, ethics is *freedom informed by reflection.*

One must understand that Foucault is celebrating an ethics of freedom, not an ethics of duty. In that respect, he looks to a future that is very different from the future faced by Kant. Foucault's future need not involve conservatively looking back into the past. Instead, Foucault's vision is focused more radically on the present. He points to the need for the formation of practices of freedom that will lead to critical resistance—resistance not merely to others, but to ourselves as well. He would certainly agree with Walter Benjamin that memorialization of the victims of history is a paradigmatic form of such resistance. When we remember to remember we are already on the way to our own future—which we know, of course, will never turn out to be what we expect.

Does this account of power and resistance thus involve a theory of temporality? Despite the widespread assumption that Foucault is largely a spatial thinker, as evidenced by his emphasis of panoptical visibility, he does indeed have well-worked-out views about temporality. For instance, in his 1973–74 lectures entitled *Psychiatric Power* he distinguishes the temporality of the power of sovereignty from that of the disciplinary power that emerges in the late eighteenth century. Whereas the power of sovereignty involves the traditional model of power as possessed by the sovereign and imposed from the top downward on the subjects, disciplinary power is more diffuse and permeates society in a capillary fashion. The sovereignty model of power looks backward to the principle that founds its authority. This principle can be divine right or blood or birth, and it is repeated discontinuously in rituals, ceremonies, and narratives that reestablish the tradition from time to time. Disciplinary power, in contrast, "looks forward to the future, towards the moment when it will keep going by itself and only a virtual supervision will be required, when discipline, consequently, will have become habit."[39] Whereas sovereignty depends on the idea of precedence, and is thus essentially connected to the past, discipline is futural, and it involves a temporal gradient aiming at the telos where discipline will function permanently of its own accord.

Individuals in modern society are caught in the pincers of these two models of power. The experience of the temporality of modern society is the effect of this pincers. Modern temporality is produced through the conflict that results when the closure that the materiality of the past would impose on the present is opened up by the processes of desubjectification.[40]

The thoroughly temporal character of Foucault's thought is difficult to see at first because it can be found in so many aspects of his work. Although as an *archaeological* or descriptive historian he concerns himself with making philosophical points by studying the past, as a *genealogical* or critical historian he writes the "history of the present." The use of the methods of archaeology and genealogy is not confined to different periods of his life. Many of his writings combine both of these methods.

Accordingly, Foucault's analysis of the temporality of not only the past and the future, but also the present in particular, is double-edged. Although his two stances on the present may seem irreconcilable, on my reading which attitude he adopts depends on which methodological viewpoint he is occupying. On the one hand, as a "methodological archaeologist" who studies the past as a series of discontinuities, he cautions against attributing too much importance to any given present. The present changes, and any given set of interests will never dominate the philosophical field forever. As an archaeologist Foucault therefore warns against thinking of the present as the crucial point of rupture, or the high point, or the moment of either completion or returning dawn. Our time is not "*the* unique or fundamental or irruptive point in history where everything is completed and begun again."[41]

As a "methodological genealogist," on the other hand, his other attitude toward the present is that the present is in fact where we are now. This point is not the truism that one can live only in the present. One can, after all, be in the present but not be at all attentive to it. One can be so focused on the past or the future that one fails to attend to the transformative possibilities that can be found only in the present. Foucault's call is to live more fully in the present, which is where the action is.

From an archaeological perspective, then, the present is no different from any other present. But from a genealogical perspective, each present is significantly different from every other present. Each present has its distinctive possibilities. Therefore, each present is the only one in which *we* can act. Foucault's method of genealogy is the key to action that is specific and transformative. The task of genealogy is to trace out "the lines of fragility in the present," that is, genealogy should try to grasp "why and how that which is might no longer be that which is."[42] The point of genealogical philosophy is to open up "a space of concrete freedom, that is, of possible transformation."[43]

As I explain in more detail in the postscript to this book, the genealogical method should be recognized as Foucault's most lasting contribution to the recent history of thought. Whatever stories are told from now on, their emplotment ought to be genealogical, however else they are also constructed. Contrary to Deleuze, for instance, genealogy is not necessarily opposed to dialectical emplotments. The reason genealogy can be dialectical is that it can serve either of several functions. For instance, genealogy may unmask aspects of ourselves that we have acquired through domination, and therefore may want to reject. Or genealogy may vindicate aspects of ourselves that we have overlooked because they are so close to us and so crucial to our identities. Perhaps genealogy can go beyond these critical dimensions of self-analysis and become more positive, even prophetic, by showing us a way out of the past that leads to a more open future. But whichever emplotment genealogy adopts—and here is the crucial difference from the Hegelian dialectic—genealogy never tries to tell a single story that is true for everybody. Genealogy does not attempt to chart a common course for all humanity, let alone all of history.

Jacques Derrida

As for Derrida's turn toward the normative and the political, in *Rogues* he denies that he started writing political philosophy only as late as the 1980s and 1990s.[44] He says that because his notion of *différance*-with-an-a always involves the political, his writing, which is always about *différance*, was therefore always political. I would say that if his early writing has political implications, this work is not as explicitly political as the later writings. To say that he took a political turn in the 1980s is not to say that he is being inconsistent. On the contrary, he is working out the potential of deconstruction's applicability in the practical sphere. Just as Foucault begins discussing ethical issues only after working out cognitive and theoretical issues, Derrida similarly begins discussing writers such as Benjamin and discovering value notions that he found to be "undeconstructible" only after working out his technical vocabulary (including such terms as *différance*, iteration, undecidability, and even deconstruction).

One such undeconstructible is justice. In studies from "The Force of Law" and *Specters of Marx* to *Rogues*, as we saw in chapter 4, he brings out what can or cannot be changed in the political sphere of social values. Derrida is insistent that deconstruction is not giving up the right to use the justificatory language of class struggle and social revolution. He is not slipping back into a neo-liberal reformism. Reform is what replaces revolution when people lose their sense of what they want. If "revolution" is Marx's way of overcoming time, then Derrida thinks that his roguish stance should not be labeled "apolitical."

As poststructuralism is superseded in the 1980s with this shift of philosophical interest toward normative issues, language and subjectivity also change positions in the pantheon of philosophical topics. Structuralism is standardly understood to have taken the linguistic turn by displacing consciousness and replacing it with language as philosophy's main preoccupation. Poststructuralism then tends to continue this disregard for subjectivity. In the phase after 1984, however, this prioritization itself is abandoned. Even the philosophers such as Foucault and Derrida who might have still been pursuing the grail of transcendental philosophy give up the effort to make one phenomenon foundational. The result is that anything can become the object of philosophical investigation. There is a renewed interest in consciousness in both the analytic and continental traditions, and there is no need to think that language is the model for understanding everything else.

Slavoj Žižek

With the death of Derrida, Žižek has emerged as perhaps the best-known living European political philosopher. When he says that poststructuralism was wrong-headed in attacking the Cartesian cogito, however, he appears to go backward rather than forward in the history of philosophy. If he restores interest in the Cartesian cogito, he does not restore it to its pride of place. Although he thinks there is something called the cogito that philosophy can analyze, he is not claiming, as far as I can see, that the cogito is axiomatic for the rest of philosophy, as it is for Descartes. This is another reason for thinking that although Žižek is interested in consciousness and subjectivity, he is not really interested in the cogito as described specifically by Descartes.

In *The Ticklish Subject* (1999) Žižek does see that Heidegger's *Kant and the Problem of Metaphysics* is an influential text. Žižek maintains that "what Heidegger actually encountered in his pursuit of *Being and Time* was the abyss of radical subjectivity announced in Kantian transcendental imagination, and he recoiled from this abyss into his [later] thought of the historicity of Being."[45] In short, because Heidegger could not work out an account of subjectivity, he gave up the transcendental enterprise altogether. The doctrine of radical subjectivity is tied to transcendental philosophy, because if the subject constitutes the world, then the world is just the expression of one person's will. That is, I believe, a fair if brief characterization of the thinking of those who want to connect Heidegger's philosophy to his politics.

But what if that from which Heidegger recoiled was not the abyss of radical subjectivity but of radical temporality? What if the topic of the Kant book is not subjectivity, but temporality, as I argue in chapter 1? There I maintain that Heidegger sees temporality as what temporalizes, and thus subjectivity is not the fundamental point, but is constituted by temporality. Boredom then illustrates more concretely how subjects are produced. Žižek notes that the findings about radical subjectivity of the Kant book are never taken up again, and that the line of thinking about Kant's transcendental imagination is a philosophical dead-end. But Heidegger did pursue the investigation of temporality in other texts, so Žižek might be coming to the right conclusion—that Heidegger did not pursue the idea of radical subjectivity any further—for the wrong reason. The right reason might well be that for Heidegger temporality is more primordial than subjectivity, and that Heidegger's writings through the 1920s are attempting to establish that thesis.

The standard interpretation of Heidegger's politics, Žižek tells us, is that Heidegger's errance into National Socialism is due to the fact that *Being and Time* is still a work of transcendental philosophy. My reading attempts to see that book and others of this period not as transcendental arguments, but as hermeneutical interpretations. Interpretations are always revisable, and thus allow for more political flexibility than transcendental arguments. If this line of interpretation of Heidegger is *plausible* (and that is the strongest claim I will make for it), then there is no connection between Heidegger's political "stupidity" (as Žižek calls it) and his philosophical acumen. I am left wondering why Žižek does not worry that his concern to reestablish the Cartesian cogito might have the same implications that he attributes to Heidegger's alleged encounter with radical subjectivity. Will not bringing back the Cartesian cogito put us back on the train of transcendental philosophy, where the next stop will be Kant all over again? Would that initial and apparently innocuous move not be likely, then, to lead to the eternal recurrence of a dubious politics?

Žižek's answer is perhaps best stated in his early work, where he resuscitates rather than buries the politics of May '68. He has a special fondness for the slogans of those times, one of which was "*Soyons réalistes! Demandons l'impossible!*"[46] In the current idiom, a good translation might be "Get real! Demand the impossible!" What this means as a political slogan is that unjust social phenomena such as the globalization of capitalism can be criticized even if there are no obvious alternatives. "Real" and "impossible" are of course quasi-technical terms for Žižek. On his account, the real is a symbolic construction that nevertheless has serious social consequences. There may be a hard kernel somewhere in our experience that is like the tiny piece of grit that eventually becomes the pearl in the oyster. Frankfurt School critical theory tries to identify this kernel by unmasking the ideology that conceals it. Critical theorists argue that social agents have real interests that get misrecognized as a result of coercive power relations. Žižek, however, maintains that the idea of real interests is itself only ever an interpretive construction, and that the conception of ideology as a representational illusion that veils social reality is to be avoided.

This denial of social reality puts Žižek in the transitional paradigm that comes after poststructuralism, and that I call "Post-Critique." For Žižek ideology is no longer to be theorized as false consciousness, or an illusory representation of the social totality. The idea of social totality emerges "only when it fails," as it necessarily will insofar as totality is "set on effacing the traces of its own impossibility."[47] Žižek takes this impossibility of representing the social totality from within as implying that we have to give up altogether the idea that the relation of appearance to reality is a representational and epistemological matter. There is thus no real social totality and the conception of the social reality is instead viewed as only an "ethical construction."[48] Any positing of the social totality is a way of dealing with "some insupportable, real, impossible kernel."[49] "Get real!" in effect tells us to "deal with" whatever our problem is. Put this way, of course, the advice does not seem particularly helpful. What I take Žižek to be saying, however, is that resistance need not be motivated by a clear vision of the ideal society. Resistance rejoins time and temporality by action in the present that is motivated by the past, even if the future is always uncertain. There is thus an ethical obligation to resist injustice even if a theory of what a completely just arrangement would be is not fully worked out.

Strategy 4   Dual Temporalization: Deleuze on Aion and Chronos

I raise the foregoing questions about Žižek not because I find that Žižek's politics are misguided, but because I have a higher regard for poststructuralism and its aftermath than he does. In my view, there is a coherent paradigm that runs through poststructuralism and after. Call it what you want—deconstructive genealogy, or normative hermeneutics, or vindicatory genealogy—but it is the enterprise that attempts to understand how temporal beings become social individuals who are not simply private subjects, but who are situated in a concrete context called the "world." To take a case in point, I proceed toward a conclusion by discussing how Deleuze as a philosopher who transcends the category of poststructuralism reconciles the contrasting phenomenological versions of temporality that Husserl and Bergson propose. I label this solution "dual temporalization."

In *Logique du sens* (1969) Deleuze identifies two different senses of what he calls time but that I think of as temporality. Using terms from the ancient Stoic tradition, he calls one "Aion" and the other "Chronos." He does not name explicitly either Husserl or Bergson in association with these two different views of temporality. Aion strikes me, however, as the more Husserlian sense of time as diachronic and linear. In contrast, Chronos emphasizes the present in a more Bergsonian manner. Chronos is time "grasped entirely as the living present in bodies which act and are acted upon."[50] Aion is time "grasped entirely as an entity infinitely divisible into past and future, and into the incorporeal effects which result from bodies, their actions and their passions."[51] For Chronos, "only the present exists in time and gathers together or absorbs the past and future."[52] For Aion, "only the past and the future inhere in time and divide each present infinitely."[53] Aion is measured by the Instant: "the entire line of the Aion is run through by the Instant which is endlessly displaced on this line and is always missing from its own place."[54] Whereas Aion is the pure empty form of time, and its present is the Now or Instant that has no thickness at all, the present of Chronos is "vast and deep."[55]

Although the reference to the "living present" could be a reference to Husserl, and would identify Husserl with Chronos, as Leonard Lawlor thinks,[56] there is a standard way of interpreting Husserl whereby Husserl is identified more with the descriptions of Aion. A typical pronouncement of this way of reading Husserl comes from Rudolf Bernet in the reference book, Blackwell's *Companion to Continental Philosophy*. Bernet interprets Husserl as saying that the present can never be perceived because it slides immediately into the past: "Consciousness never coincides with its present lived experience. Strictly speaking, there is no such thing as present consciousness."[57] This interpretation suggests to me not the vast and deep present of Chronos, but the empty form of the Instant as well as the events that are, as Deleuze says, "not living presents, but infinitives: the unlimited Aion, the becoming which divides itself infinitely in past and future and always eludes the present."[58]

Furthermore, when Deleuze contrasts phenomenology and Bergson in *Cinema 1*, he draws on Sartre to say that the phenomenologists such as Husserl and Sartre conceive of the intentionality of consciousness as the beam of a flashlight. In contrast to this notion of intentionality, whereby consciousness is always "consciousness *of* something," for Bergson the opposite is the case: "Things are luminous by themselves without anything illuminating them: all consciousness *is* something, it is indistinguishable from the thing, that is from the image of light."[59] Bergson's conception of ontology is thus more in the ancient tradition than in the modern one, and aligns more with Chronos than with Aion.

Or perhaps Deleuze identifies Bergson not with Chronos, but with Cronos—a different deity. As seen in the crystalline view of time, Deleuze says in *Cinema 2* that Cronos is "the perpetual foundation of time, non-*chrono*-logical time."[60] The connection to Bergson is implied when Deleuze echoes Bergson's vitalism and says that Cronos is the powerful "Life" that grips the world.[61] This identification is then underscored by Deleuze's summary of Bergson's major theses on time: "the past coexists with the present that it has been; the past is preserved in itself, as past in general (non-*chrono*logical); at each moment time splits itself into present and past, present that passes and past which is preserved."[62] Although the view commonly attributed to Bergsonism is that "duration is subjective, and constitutes our internal life," Deleuze thinks that, on the contrary, for the real Bergson "the only subjectivity is time, non-*chrono*logical time grasped in its foundation, and it is we who are internal to time, not the other way round."[63] For both Bergson and Proust, says Deleuze,

Time is not the interior in us, but just the opposite, the interiority in which we are, in which we move, live, and change. . . . In the novel, it is Proust who says that time is not internal to us, but that we are internal to time, which divides itself in two, which loses itself and discovers itself in itself, which makes the present pass and the past be preserved.[64]

At this important intertextual juncture we can see that Deleuze's reading of Bergson is the latest attempt in the history of temporality to fill out Heidegger's reading of Kant. As we saw in chapter 1, Heidegger inverts the standard reading that takes Kant to be saying that the mind imposes time on experience. When Deleuze says "subjectivity is never ours, it is time,"[65] I understand him to be giving us a Bergsonian version of Heidegger's *Kantbuch* whereby it is not the subject that temporalizes, but temporality itself that temporalizes. Insofar as Deleuze affirms that time is not in us, but that temporality is an auto-affection, he is indirectly repeating and expanding on Heidegger's claim that it is not subjects that produce temporality, but that temporality produces itself and subjectivity follows. Once again, temporality is seen as the Ur-phenomenon that makes subjectivity possible.

To continue the interpretation of Aion and Chronos, even if these passages associate Bergson with Cronos rather than with Chronos per se, Cronos, "the powerful 'Life' that grips the world," is still a variant of Chronos, the "vast and deep" living present. This association confirms the connection of Chronos more with Bergsonism than with phenomenology, at least on one standard reading of Husserl described above. Assuming that Cronos is a variant of Chronos, and thus that Deleuze may have some Bergsonian theses in mind when he discusses Chronos, we can reasonably ask how the concepts of Aion and Chronos can be reconciled. From one perspective, that of traditional logic and reason, they cannot be reconciled. Deleuze admits that they would standardly be viewed as "mutually exclusive."[66] He remarks,

We have seen that past, present, and future were not at all three parts of a single temporality, but that they rather formed two readings of time, each one of which is complete and excludes the other: on the one hand, the always limited present, which measures the action of bodies as causes and the state of their mixtures in depth (Chronos); on the other, the essentially unlimited past and future, which gather incorporeal events, at the surface, as effects (Aion).[67]

From another perspective, however, that of a philosophy of difference and of becoming, Chronos and Aion are two different but complementary ways of grasping temporality. Chronos is the concurrent, living present, whereas Aion represents the perception of time as stretched out and divided infinitely into past and future. In this sense, they would be the two sides that any theory of temporality would have to take into account. From this perspective, Chronos and Aion together could be mapped onto both Husserl's and Bergson's accounts. Deleuze says Chronos and Aion represent "two simultaneous readings of time."[68] Expressed this way, the aura of paradox disappears, as there is nothing mysterious about viewing temporality both synchronically and diachronically, or transversally and longitudinally. Husserlians imagine temporality as flowing along in a stream. Bergsonians take a transverse view across the flow and capture the present. The two ways of diagramming temporality as either a linear graph or an expandable cone are not necessarily in conflict. In fact, both express essential features of temporality.

In the scission at the heart of this dual temporalization, that is, in the nonspatial divide between these two readings of temporality, the ideal game, the ultimate game, takes place. I read this notion of the ideal game as a metaphor for living a life. For Deleuze, the ideal game is played "without rules, with neither winner nor loser, without responsibility, a game of innocence, a caucus-race, in which skill and chance are no longer distinguishable."[69] These features—no rules, no winners or losers, no responsibility—form a list of attributes that Deleuze calls nomadic. These attributes also represent the normative ideals of poststructuralism, for both its critics and its proponents. Deleuze's philosophy of difference embodies countercultural norms that are rejected by present-day theorists as different from one another as Habermas and Žižek. Deleuze's philosophy of becoming insists on chance as opposed to necessity. The necessity to which he is opposed includes not only causal necessity, but dialectical necessity as well. For Deleuze there is no higher synthesis of these two conflicting accounts of time in which they are mutually rejoined into one. Instead, they must be repeated in their difference with each move of the "game."

Each move of the game of life makes up its own rules, and the rules can always be changed. The game is also chancy, and a paradigm of the ideal game is not the philosopher Wittgenstein's favorite philosophical example, chess, but the poet Mallarmé's game of throwing dice. (The title of Mallarmé's most famous poem is "Un coup de dés jamais n'abolira le hasard.") Giorgio Agamben notes that at the beginning of the history of philosophy Heraclitus figures Aion as a child playing with dice.[70] Furthermore, Nietzsche emphasizes the issue of chance and necessity by way of "the great dice game of existence."[71] Deleuze says of Nietzsche (but also, I take it, of himself):

Nietzsche identifies chance with multiplicity, with fragments, with parts, with chaos: the chaos of the dice that are shaken and thrown. *Nietzsche turns chance into an affirmation.* . . . What Nietzsche calls *necessity* (destiny) is thus never the abolition but rather the combination of chance itself.[72]

Of course, there is no way of winning the ideal game of life. The very act of playing is itself an affirmation of chance. In Bergsonian terminology, Mallarmé's poem is a highly contracted point in comparison to Proust's novel, which could be viewed as aiming at the most relaxation possible, perhaps to hold off the movement toward the future, and the end. But of course it is an illusion that the longer the read, the longer the time is that remains to read it. Time waits for no one, no matter how long the book is.

Closing Time

If the way this book has mapped out the terrain of temporality is on the right track, it helps to clarify the phenomenological features of the time of our lives. Even if temporality and time are not really separable, distinguishing them conceptually prevents anyone from closing the door on the possibility of understanding time for anyone else who is not trained in contemporary physics or who has issues with metaphysics. By starting from "the time of our lives," rather than from "the time of the universe," nonphysicists and antimetaphysicians can investigate the crucial phenomenological and existential issues that motivate many to start thinking philosophically in the first place. Rather than close the philosophical door to an understanding of time by restricting the analysis to the nature of objective time—the time of the universe—phenomenologically the door is opened to a dimension of temporality that is real, insofar as it characterizes the temporality of our lives, even if temporality is not the same as either the objective time of the physical universe or the subjective sense of time passing within consciousness. Closing "time" opens the way to "temporality."

Awakening the phenomenological sense of temporality is admittedly a process that can take a long time, and the process is not identical to that of accessing subjective time through psychological introspection. One respect in which phenomenology differs from psychological introspection is that it is based not only on observations of self, but also on observations of others, including their agreement to the results of the analysis. The analysis, furthermore, does not involve simply telling one's own story from an internal point of view, but it depends on giving systematic arguments and generating a consistent account of the human being.

To sum up these strategies for reconciling the sting of time and the enjoyment of time, the basic tension is between the sense that we always have some time remaining and the sense that time is running out. The initial set of strategies discussed in this chapter tries to reconcile these two conflicting senses through reminiscence and remembrance. The first of these was achieved through the joy that Proust discovered, and that we discover through Proust. Whereas that joy is prereflective and cannot be willed, Benjamin's strategy of remembering against the grain reconciles the present to the process of becoming past by establishing a sense of justice.

Rather than justify the present by looking ahead, Benjamin's allegory suggests that the present can only justify itself by rectifying past injustices. Joy and justice are thus the signs that time and temporality have been integrated again.

Second, the hermeneutic tradition of Husserl, Heidegger, and Gadamer also attempts to reconcile the present and the past. Husserl's notion of retention, Heidegger's concept of resoluteness, and Gadamer's *wirkungsgeschichtliches Bewusstsein* represent different ways of becoming conscious of the power of the past over any interpretation of what is essential in the present. Resoluteness is the means by which we face a finite future. Time and temporality are thus integrated in a realistic acceptance and anticipation of our finitude.

The third strategy of reconciliation involves the politics of the future. The question of politics is, what social arrangements can we all agree to, or can we actively try to bring about? Moreover, *how* can we align ourselves politically, given that there is no standpoint from which we can see which social arrangement is the best? A politics of Refusal and an attitude of either passive aggressivity or aggressive passivity may reflect only angry *ressentiment* of the darkness of the future.

A more considered and reflective attitude is projected by the fourth set of strategies. The synthesis of the previous three attitudes can potentially be achieved by a temporalization that combines a forward-looking attitude that is fully informed both by sympathy for those who suffered past injustices as well as by a practical sense for present possibilities. For Deleuze, of course, reconciliation is not a synthesis in the sense of a merger or a unification of the two senses. For him repetition is of an essential difference that is preserved not in the concept, but in the phenomenon, which in this case is lived temporality. Whatever the prospects for such a synthesis are, for now we can conclude with an optimistic thought, namely, that if temporality temporalizes, then it is open to us to temporalize in the way that best brings about both joy and justice.

Postscript on Method: Genealogy, Phenomenology, Critical Theory

This book has studied temporality as seen through the different lenses of a variety of philosophical approaches. If phenomenology has been the central method under investigation, other methods have included Bergsonism, critical theory, pragmatism, deconstruction, and hermeneutics. This study has allegiances with all these traditions, but it understands itself in particular as a form of the critical history of philosophy, one that employs genealogical strategies. This postscript will pull together the threads of discussion of the three philosophical methods that are most at stake in this book, namely, genealogy, phenomenology, and critical theory. In particular it will focus on the question, "What is genealogy?" The intention is to clarify how genealogy functions as a method of critique by contrasting it with these other conceptions of philosophical method. I suggest in conclusion how my project in this book, as well as the project of "critical history" more generally, can be understood in terms of these rubrics and related distinctions.

To start with a tentative working definition, genealogy is a philosophical method of analysis of how certain cognitive structures, moral categories, or social practices have come into being historically in ways that are contrary to the ordinary understanding of them. In continental philosophy genealogy is a method often ascribed to the poststructuralist philosophers. These philosophers inherit their method from Nietzsche and wield it against other dominant trends in French philosophy. One of these trends is the phenomenological tradition inspired by Edmund Husserl and carried out by Maurice Merleau-Ponty. Another is the Hegelian or dialectical method that forms the basis of critical theory and that was stimulated in France by Alexander Kojève's Paris lectures on Hegel from 1933 to 1939. The movement of Bergsonism was in decline in the second half of the twentieth century, but I have argued that its revival by Gilles Deleuze keeps it in contention.

Whereas genealogy can be attributed to several recent French thinkers including Michel Foucault, Gilles Deleuze, and even Jacques Derrida—the attribution becoming more controversial in the order in which I have named them—the term "poststructuralism" could well be questioned. Although these philosophers are often grouped together under the label of poststructuralism, in fact that label says nothing about what they have in common. The most that this label does is gesture toward whatever comes after structuralism. There was, however, never really any structuralist *philosophy.* The famous structuralists were anthropologists, linguists, or psychoanalysts. Furthermore, the styles of the poststructuralist philosophers are so different from one another that they can just as easily be pitted against one another as allied under such a vacuous term as poststructuralism.

Unlike the parochial term "poststructuralism," "genealogy" has been adopted as the name for a distinctive method by a variety of philosophers in both the analytic and the continental traditions. Nietzsche, who is usually credited with the initial use of the genealogical method, in fact attributes it to earlier British philosophers. In a previous paper I argued that one of these must be David Hume.[1] Resemblances and connections should not obscure, of course, the significant differences between Hume's and Nietzsche's employment of genealogy. As Nietzsche understands that difference, Hume and the other British genealogists dig under psychological phenomena to identify the shared features that run through experience. Hume's use of genealogy thereby vindicates standard morality. In *Truth and Truthfulness* Bernard Williams thus calls this usage "vindicatory genealogy."[2] He sees Nietzsche's genealogy, in contrast with Hume's vindicatory genealogy, as an "unmasking" method that explains how morality emerged from nonmoral and even antimoral forces. I will use Williams's distinction between vindicatory and unmasking philosophy in the following discussion of how genealogy functions as critique.

By calling morality into question, Nietzsche's unmasking genealogy attacks vindicatory genealogy as well. Genealogy that is thoroughly unmasking will challenge everything, including itself. Just as Nietzsche continually questions his own questioning, genealogists have to risk regress by asking whether their own views are not simply perspectives on perspectives. Genealogy thus becomes a methodological challenge to the rationality and coherence of its own interpretations of self and world. I hasten to add, however, that genealogy need not thereby abandon its own interpretations. Doubting is not the same as denying. If the genealogy finds no grounds for suspecting its own rationality, it can assume that its understanding of the phenomena in question is sound, at least for the time being. Vindicatory genealogy can thus survive the attack by the unmasking type of genealogy.

Of the principal rivals to genealogy, namely Bergsonism, phenomenology, dialectics, and critical theory, I suggest that the first two are vindicatory and the second two belong to the unmasking type of philosophy. To explain this distinction, I begin with Bergsonism and then I turn to dialectical critical theory before coming to the relation of genealogy and phenomenology.

In reflecting on Henri Bergson, as we see in chapter 3 of the present study, the phenomenologist Maurice Merleau-Ponty lists three "Bergsonian" doctrines.[3] The first is that intuition is prior to intellect and logic. The second is that spirit has primacy over matter. The third is that life (or vitality) is more primordial than mechanism. While Merleau-Ponty insists that these doctrines are merely popularizations of Bergson's real philosophy, the appeal to intuition brings out the unsuspicious character of Bergson's philosophy. Because intuition is in primordial contact with things, intuition is the ultimate arbiter. One could argue that Merleau-Ponty is assimilating Bergson to his own conception of philosophy as phenomenology. For Merleau-Ponty, phenomenology "tries to give a direct description of our experience as it is."[4] If metaphysical reflection distorts immediate experience, phenomenology aims at "re-achieving a direct and primitive contact with the world"[5] and "re-learning to look at the world."[6] This characterization of phenomenology makes it vindicatory. The idea is that prereflectively we have a basic relation to the world that is distorted by ordinary introspection and reflection. Phenomenology is a more rigorous way of attending to prereflective experience. Through phenomenology we can vindicate philosophically what we already understand about experience, even if "understanding" is not equivalent to explicitly knowing. For similar reasons, the recapture of intuition in Bergsonian philosophy would also be vindicatory. We are finding out the truth that we already understand through intuition, and philosophy is the systematic articulation of this more intuitive knowledge.

The situation with dialectics and critical theory is markedly different. These methods are more unmasking than vindicatory. From the dialectical perspective the reliance of both Bergsonism and phenomenology on appeals to intuition resembles what is called in the continental tradition the "myth of presence" and in the analytic tradition the "myth of the given." That is, there is no sensory givenness or immediate presence that is not already permeated by conceptual or linguistic factors. Also, vindication tends to overlook the phenomenon of meaning change. In the Kantian tradition the meaning of concepts stays the same over time. Arranging concepts coherently in a comprehensive system is a Kantian way to vindicate these concepts. In contrast, in the tradition of Hegel's *Phenomenology of Spirit*, there is no immediate given, and the concepts change their meaning as they are combined with other concepts. As Richard Rorty remarks, "it is much easier to formulate specific 'philosophical problems' if, with Kant, you think that there are concepts which stay fixed regardless of historical change rather than, with Hegel, that concepts change as history moves along. Hegelian historicism and the idea that the philosopher's job is to draw out the meanings of our statements cannot easily be reconciled."[7]

Meaning change is said by its advocates to be holistic. Holism holds that some concepts cannot remain the same while others change. Instead, change in some concepts results in changes in all. Nevertheless, concepts can change at different rates. Foucault notes, for instance, in his study of the history of ethics that the moral rules for sexual conduct have varied relatively little since the ancient Greeks. Where there has been genuine change is at the deeper level of what he calls "ethical substance"—that is, the underlying self-understanding that explains why one wants to obey the moral rules and what one thereby hopes to become.

This meaning change thus makes it not only possible, but also highly probable that the present understanding of any given particular idea will be decidedly different from earlier understandings of it. Nietzsche's genealogy of morality claims, for instance, that the term "good" changes its meaning from when it was paired with "bad" to when it begins to be contrasted with "evil." We can still hear this meaning change when we use "good" not as an expression of natural aesthetics, as when we say that something tastes good, but as a moral term. The *moral* idea of the good is normative in a different sense for Nietzsche. Moral goodness is not to be vindicated in Hume's fashion, but is instead to be unmasked genealogically. On the Nietzschean analysis, the moral notion of goodness is a double negation and an abstraction. The double negation results first when something that is good in the natural sense is turned into something evil in the moral sense. Nietzsche then concludes that "good" in the moral sense means whatever is not evil. "Good" in the moral sense is thus twice a negation and morality accordingly becomes a string of largely negative commands: "Don't do this, don't do that!"

Genealogy and Critical Theory

What genealogy reveals about morality can be generalized to apply to our social interactions and institutions. We begin to suspect that social arrangements that we take to be just are in fact hiding oppressive relations of force. How could we substantiate these suspicions? How does one know that one's understanding of the social world is the distorted result of oppressive social conditioning and not the way things really are? The method that has perhaps the most to say about this problem of social and political thought is critical theory.

Critical theory differs from traditional theory, Max Horkheimer tells us in a 1937 essay, in being perspectivistic and self-referential rather than aspiring to neutrality and objectivity.[8] Critical theory is perspectivistic because in social theory, the "reality" that is in question is social. In a divided society, furthermore, no single perspective can claim to be exclusively correct. A form of resistance to cooptation by the dominant class, critical theory takes up the perspective of the oppressed. Horkheimer writes that the purpose of critical theory is not "the better functioning of any element in the [social] structure. On the contrary, it is suspicious of the very categories of better, useful, appropriate, productive, and valuable as these are understood in the present order."[9] There is thus a decidedly deconstructive side to critical theory.

Critical theory is also self-referential. Traditional theory assumes that society is the way the theory tells us it is, and that its own role does not need to be explained. Critical theory, in contrast, examines its own place in the social context. If the society is divided, then all perspectives are biased, and critical theory has to be self-critical and alert to its own distortions. Traditional theory maintains that all presuppositions of a theory must be brought to light and examined before a theory can be considered to be sound. Critical theory is pointing out that insofar as traditional theory does not investigate itself and its own social role, traditional theory is criticizable on its own grounds for leaving presuppositions unexamined. Critical theory, in turn, "requires for its validity an accompanying concrete awareness of its own limitations."[10]

Insofar as critical theory is perspectivist, it is a form of situated or standpoint knowledge. Critical theory starts from specific social problems and works toward an explanation of how they came about. It also takes into account the question of who is speaking, or which social standpoint is at stake. In this respect, critical theory and genealogy are closely allied. They both are what Foucault refers to as "history of the present" and what Horkheimer calls "criticism of the present."[11] In Foucault's terms, they both offer explanations of the emergence of current "problematizations." A problematization for Foucault is the way some issue becomes a normative issue, even if it is never resolvable into a "just and definitive solution."[12]

If particular problematizations were to disappear, the critical theory itself would no longer be needed. In this respect, critical theory is unlike traditional theory, which assumes that not only the truths it discovers but also the problems it investigates are eternal. Critical theory and genealogy are both more historical than traditional theory. They can claim validity only so long as they are useful. Their goal is not scientific objectivity so much as social improvement.

Both critical theory and genealogy have a similar attitude toward the past, present, and future. In relation to the past, their recognition of the situatedness of the inquiry means that the so-called genetic fallacy is no longer considered fallacious. Traditional theory assumes that the question of how we acquired our beliefs and knowledge is irrelevant to the validity of those beliefs. In contrast, both critical theory and genealogy are asking precisely how we came to forget the contingency of the historical beginnings of our practices and why we persuaded ourselves that these practices were necessary and universal rather than arbitrary and contingent. In relation to the present, both critical theory and genealogy view the theory itself as part of the problem, such that if there were no problem, there would be no need for the theory. In relation to the future, they both aim at social transformation, not justification of current social arrangements, which are veiled power relations.

Critical theory and genealogy both try to unmask power and show it for what it is. Insofar as this unmasking works, it does not necessarily bring about social change, although it does make social transformation more likely. Horkheimer remarks aptly about the social sphere, "[human beings] can change reality, and the necessary conditions for such change already exist."[13] Genealogy recognizes more cautiously that it does not change the world, but it does prepare the world for change. By disrupting the fatalism resulting from resignation to the inevitability of oppressive social institutions, genealogy frees us for social transformation, even if it does not tell us precisely what to do or where to go.

If this way of explaining the kinship of genealogy and critical theory is right, it should not obscure some crucial differences between them. Critical theory in the form it takes in the Frankfurt School relies on two notions that are not to be found in genealogy, at least in the form that genealogy takes for the French poststructuralists. The first involves the philosophy of history. The second is the appeal to real interests. To explain the first point, critical theory needs some point of purchase to explain its belief in the correctness of its own findings. In Horkheimer's 1937 essay he draws on a philosophy of history—one that Martin Jay attributes to Rosa Luxembourg—to justify critical theory's aspirations to social amelioration.[14] Horkheimer sees his current society as the result of either oppression or the blind outcome of competing forces, but "not the result of conscious spontaneity on the part of free individuals."[15] Critical theory aims at the ideal of a free society in which humanity can become self-aware with no oppression or exploitation.[16] The idea is that if critical theory promotes the historical progress of human freedom, then it can claim to be emancipatory—in contrast with traditional theory, which fixates on the status quo and thus serves to preserve the unequal power relations of modern society.

When Horkheimer teams up with Adorno later in the 1940s to write the *Dialectic of Enlightenment*, he comes to share Adorno's more pessimistic vision of society. A regressive story is just as teleological as a progressive story, however, and genealogy is opposed to any such teleological philosophy of history. Whereas normally universal history, or the history of everything, combines both an eschatological and a teleological view of historical progress, Derrida distinguishes these. What he argues for, as we saw in chapter 4, is "messianicity without messianism." In other words, he rejects a teleological view of the future as predictable and progressive. In contrast, the eschatological future is always unpredictable and unexpected. Breakthroughs can happen at any time. Eschatological messianicity looks to the future and calls for present action without delay. Neither optimistic nor pessimistic, the notion is intended to explain how both optimism and pessimism are possible in the first place. "*Anything but Utopian*," says Derrida, "messianicity mandates that we interrupt the ordinary course of things, time and history *here-now*; it is inseparable from an affirmation of otherness and justice."[17]

Foucault similarly rejects the idea of progress because it implies a standpoint outside or above history from which to make the judgment that history is progressing and universal freedom is increasing. There is no standpoint from which to celebrate an optimistic utopianism. Foucault's own rhetoric, however, has perhaps not been as consistent on the question of the regressive slide into barbarism. Foucault has been interpreted by sensitive readers, including Clifford Geertz, as telling a dystopian story of the rise of unreason.[18] *Discipline and Punish* in some places does read as if disciplinary power has spread insidiously until it threatens to consume the entire society. The vision of the penitentiary system turning into the completely carceral society is a powerful one that Foucault is not really entitled to use. That he recognizes this limitation becomes clear, however, insofar as if the society were totally carceral, there would be no room for a subversive book such as *Discipline and Punish* itself.

The second point of difference with critical theory concerns Foucault's abstention from use of either the idea of real interests or the term "ideology." The Frankfurt School relies on the Marxian-Lukácsian idea of false consciousness as a way of explaining why people act contrary to what is obviously in their real interests. Horkheimer invokes the concept of ideology to explain what holds together social structures that would otherwise collapse,[19] and he sees ideology as permeating *every* social stratum.[20] Why, for example, would people in a company town allow a factory to build a large, highly polluting smokestack in their midst except for the fact that their livelihood depends on the success of the company? Cases such as this one illustrate how people misperceive their true preferences as a result of social coercion. Foucault abstains from the idea of both true interests and false consciousness for two reasons. First, he thinks that it is inappropriate to speak of false consciousness if there is no true consciousness. Second, he thinks that true consciousness would require theory to have a "view from nowhere," that is, a point of view outside society that would be necessary for a history of everything. But such a standpoint is impossible.

In sum, genealogy is more resistant than critical theory to the positing of a bottom line for social criticism. Who is to say, after all, why people perceive their interests one way rather than another? Theory lacks the grounds for identifying some interests as true or real and others as false or illusory. Genealogy strikes me as being more thoroughgoing than critical theory and *Ideologiekritik* in that it challenges the very idea of ideology.[21] A suspicious genealogy cannot leave anything unexamined, including itself.

Universalism

The story of critical theory cannot consider itself complete without at least some discussion of Jürgen Habermas, especially during the days in the 1980s when he saw the poststructuralist movement as being dangerously relativistic. Habermas agrees with Foucault, surprisingly enough, that there is no view from nowhere, that is, no point of view above society from which to judge the whole of society and history. Instead, he thinks that criticism is only possible from a standpoint *inside* history and society. He therefore posits the telos or ideal of the universal consensus that he believes we are invoking counterfactually every time we engage in debate and inquiry.

Many in both the analytic and the continental traditions resist this universalism of the early Habermas. Foucault, for instance, maintains in reply that social hope comes not from maximizing consensuality, but from minimizing nonconsensuality. He worries that advocating consensuality in Habermas's fashion will lead to an intolerance of difference. On the analytic side, Alasdair MacIntyre argues against Habermas's view that "allegiance to one specific set of ideal norms is a necessary condition for acts of communication" as follows:

All that writer and reader must presuppose is enough of logical, ontological, and evaluative commitment—and the commitments of the one need not be in all respects the same as those of the other—to ensure the continuities and fixed identities and differences without which each cannot by his or her own standards, even if not yet or not at all by those of the other, convict that other of inconsistency, falsity, and failure of reference.[22]

In other words, we can engage communicatively with one another without the supposition of a universal consensus. Local commitments alone will be enough to ensure that rational communication is taking place. We do not criticize all standpoints other than our own from the ideal standpoint of a metacommunity. Rather, we criticize one standpoint from other standpoints either within the current situation, or by contrast to a past that is no longer recoverable (for instance, Greek ethics or the samurai honor code).

On my reading of Derrida's development, in the early days he would have agreed with this criticism of Habermas. In his later writings, however, his methodological self-understanding becomes increasingly genealogical. In the last decade or so of his life, he is more prone to take on social, political, and ethical issues, and he begins to call his method "deconstructive genealogy."[23] His approach is appropriately called *genealogy* for two reasons: (1) because it shows that what is taken as natural and necessary is really contingent and historical, and (2) because it also makes social transformation more likely. What makes his genealogy *deconstructive* is that (1) it challenges the self-certainty of the critical attitude that thinks it is in the know, and (2) it also challenges the self-certainty of the social theorist who insists on knowing where we are to go before deciding whether to act.

Derrida thus fulfills the task of genealogy of reflecting on itself. Derrida does this by being critical of the very idea of critique. He views as too limited any critique that is merely negative with no positive conception of how to change society. At the same time, he is opposed to "methodological smugness," or "good conscience." This is the attitude of complete self-confidence of the critic who has no doubts about where society is headed. Derrida thinks that the appeal to critique is thus limited. If this methodological smugness blocks us from seeing other possibilities, then it is to be rejected as well. Critique must always be open to self-criticism.

Genealogy need not be opposed to universals. The problem is not universals per se. Although genealogy may be suspicious of claims to universality, it need not reject all appeals to universal structures or values. Thus, Foucault and Derrida both imply that Habermas's criticism of them for being relativists or nihilists is off-target. Foucault says, for instance, that although experiences are always singular, "Singular forms of experience may perfectly well harbor universal structures."[24] He then clarifies this by adding, "That [thought] should have this historicity does not mean it is deprived of all universal form, but instead that the putting into play of these universal forms is itself historical."[25] So a universal structure has a beginning in historical time, and thus a foreseeable end.

Foucault has a nuanced attitude toward the universal. He opposes universalist strategists who regard a particular injustice as inconsequential when viewed from the perspective of the greater necessity of the whole. Instead, Foucault insists that his own theoretical ethics and politics is the opposite of these suprahistorical universalists. "My theoretical ethic," he says, "is 'antistrategic': to be respectful when a singularity revolts, intransigent as soon as *power violates the universal*."[26] In other words, as a grassroots activist, Foucault will not hesitate to condemn an act that violates his sense of justice.

The method of study of universals will change for the genealogist, however. Genealogy entails a methodological nominalism whereby, in Foucault's words, "universals do not exist."[27] What does Foucault mean by universals here? Some examples are state, society, sovereignty, subjects, and madness. Foucault's "critical history" of thought maintains a studied skepticism toward these "anthropological universals."[28] Instead, the critical historian must try to see these anthropological universals as "historical constructs," as what is produced when the subject tries to make itself into its own object.

This methodological nominalism does not entail that universals do not have real effects. In contrast to phenomenology, which says that universals exist even if they are not a "thing," genealogy acts as if universals do not exist but with the caveat that they are not "nothing."[29] From a fictitious relation, Foucault maintains, a real subjection can be born.[30] Genealogy is thus the study of the birth of universals and their transformation into principles of domination. Not a form of universal history, in Foucault's hands genealogy is nevertheless a method for investigating the history of universals.

In the case of Derrida, although early on he seems to be an ethical pluralist, in his later writings he comes closer to Habermas insofar as he also insists on the social value of consensus and universal principles. Derrida projects a list of duties including those of "respecting differences, idioms, minorities, singularities, but also the universality of formal law, the desire for translation, agreement, and univocity, the law of the majority, opposition to racism, nationalism, and xenophobia."[31] Add to this list the universals that he considers "undeconstructible" and thus "unconditional," such as justice, otherness, and messianicity, and it is impossible to consider the later, genealogical Derrida a nihilist. He sums his view up in a decidedly straightforward way so that he will not be misunderstood again when he says, "I have, the unique 'I' has, the responsibility of testifying for universality."[32]

A more pressing problem for both philosophy and politics today is not the absence of universality, but the actuality of too many different claims to universality. Much of the turmoil of current politics is due to the simultaneous existence of "multiple universalities," and especially of "competing universalities."[33] As I understand these terms, a "universality" is a set of concepts or principles that is asserted to apply to everyone, everywhere, and always. When two or more of these sets occur, they are "multiple universalities." When more than one would be impossible for the one in question to accept, then these are "competing universalities." Competing views that claim universality—for instance, theologies that insist on only one true religion—may often not be able to tolerate rivals and, as history shows, may even take violent action against one another.

Genealogy and Phenomenology, Redux

Between the pair, genealogy and critical theory, I have argued that genealogy is a more consistent position than critical theory, and it has widespread appeal especially in the arena of social and political philosophy. What about the other approaches I identified? Bergsonism and dialectics are no longer in the forefront of philosophy, even if they have their proponents. Phenomenology is therefore the major competitor to genealogy. But what is phenomenology? Those who are not squarely in the field will often use the term "phenomenology" as a synonym for continental philosophy as a whole. Earlier in this postscript I gave a characterization of phenomenology based on Merleau-Ponty. The important point about phenomenology is that it thinks of itself as purely descriptive. Bringing any presuppositions about what ought to be the case into the inquiry will presumably distort the phenomenon in question. If you reflect on experience in the right way, the idea is that you could never be wrong about it. Phenomenology thus maintains that the experience is *immanent* in reflection. In contrast to genealogy, which is always suspicious of the way experience appears to reflection, phenomenology sees the experience as being available, when reflected on properly, "completely, directly, and all at once."[34]

Given this account of the phenomenological method as a philosophical program, if we follow Williams's division of philosophy into the vindicatory and the unmasking types, genealogy appears to be primarily unmasking whereas phenomenology is primarily vindicatory. These two philosophical programs would then be the principal contenders in the current arena of methodological debate. What makes this characterization too simple, however, is the question of whether there is not some middle ground between these two poles. Genealogy is often descriptive as well as prescriptive, and therefore has a phenomenological dimension to it. In turn, phenomenology can be evaluative as well as descriptive, and social as well as personal. In an essay entitled "Naturalism and Genealogy," Bernard Williams describes the account of *ressentiment* in Nietzsche's genealogy of morals as a "phenomenological representation."[35] *Ressentiment* is more than resentment or rancor. Resentment is experienced explicitly as a reaction to something or someone. *Ressentiment* lies behind an attitude, usually as its opposite. As Alasdair MacIntyre sums up the point, Nietzsche reveals how "a concern for purity and impurity provided a disguise for malice and hate."[36] Another example would be how a sincere concern for social equality can mask a fear that others will end up with more than one has oneself. We therefore have two examples of genealogical unmasking: religious love masks hate, and democratic equality masks envy.

Each of these unmaskings relies, however, on a moment of phenomenological representation. It is important to emphasize Williams's observation that the ascription of *ressentiment* is a phenomenological claim and not a psychological one. The claim is not simply psychological because the agent would not recognize the explanation. Rather, it is an analysis of what is happening under the surface, and the explanation involves factors that are not simply personal but *social*. Williams remarks,

> If this were a psychological process, it should be recognizable in an individual. But an actual process that led to the actual explanandum could not happen in an individual, since the outcome consists of socially legitimated beliefs, and they could not be merely the sum of individual fantasies. Rather, this is a social process. . . .[37]

Williams thus extends the scope of phenomenology beyond the concern with individual consciousness, subjectivity, and selfhood into the social sphere. If this ability to deal with socially legitimated beliefs that cannot be reduced to individual psychology is ascribed to phenomenology, then the boundary between phenomenology and genealogy becomes less sharp.

Although Williams suspects that not all genealogies need this phenomenological moment, he thinks that when they do include it, the phenomenology prevents anyone from going back to the earlier state of affairs. An example could be drawn further from Nietzsche's analysis of *ressentiment*. Once the good–evil distinction has taken hold, it becomes impossible to go back exclusively to the earlier

nonmoral distinction between good and bad. Once moral evaluation is in place as a social practice, a society without it is unimagi-
nable. This point does not undermine genealogy as an unmasking
critique. Recognizing the value of the phenomenological moment
does not lead to a vindication of morality so much as it makes
moral condemnation a much less absolutist enterprise. Phenomenology can thus invoke prescriptive possibilities that give it a critical dimension. For example, Heidegger's notion of authenticity
springs from a phenomenological analysis of Angst, yet it clearly
has normative implications for how to live one's life. Authenticity
is not only the phenomenological recognition of the inevitability
of death; it also involves criticism of a life in which human finitude
is ignored or denied.

Recognizing the phenomenological moment within genealogy
can be a useful way to deal with a criticism of genealogy raised by
Alasdair MacIntyre. MacIntyre's criticism involves catching the
genealogist in the fallacy of self-exemption. The genealogist's
claims, in other words, could not be true of the genealogist him- or
herself. As MacIntyre puts it, "the intelligibility of genealogy
requires beliefs and allegiances of a kind precluded by the genealogical stance."[38] In particular, the genealogical conception of the
self as always masked raises the question whether the genealogist
is similarly masked. MacIntyre thinks that the genealogist is saying
on the one hand that the self is "nothing but a sequence of strategies
of masking and unmasking," and yet, at the same time,

the genealogist has to ascribe to the genealogical self a continuity of
deliberate purpose and a commitment to that purpose which can only be
ascribed to a self not to be dissolved into masks and moments, a self which
cannot but be conceived as more and other than its disguises and conceal-
ments and negotiations, a self which just insofar as it can adopt alternative
perspectives is itself not perspectival, but persistent and substantial. Make
of the genealogist's self nothing but what genealogy makes of it, and that
self is dissolved to the point at which there is no longer a continuous
genealogical project.[39]

The metaphor of masks thus itself conceals the harder question: who is behind the masks?

This objection raises many issues and its rebuttal will require the treatise on subjectivity and self-consciousness that follows this treatise on temporality and time-consciousness. Let me suggest for now, however, that Foucault was aware of this type of objection and that he took measures late in his career to deflect it. These measures appear to take him back in the direction of the phenomenology that he had repudiated at the beginning of his career. Whereas the genealogy of power of the 1970s is a third-person standpoint, showing how subliminal power relations socially construct subjects, Foucault then turns in the early 1980s toward ethics, where individuals are responsible for the constitution of themselves, a process often referred to as "self-fashioning." Foucault suggests that there is room within any given social context for free choice, and that agents could freely constitute if not their entire selves, at least central aspects of themselves. Even in *Discipline and Punish* agents are not zombies, but they consciously permit the power relations to be exercised on themselves. Although prisoners may not have much freedom to resist the penitentiary schedule that is intended to reform them, soldiers, students, and workers do have the capability of applying power relations to themselves. Or at least they can be aware that they are submitting themselves to disciplinary procedures.

MacIntyre's criticism fails to note that Foucault has much to say about the first-person standpoint of self and subjectivity. The phenomenological self may be made up of the interaction of various forces, as the Nietzscheans would have it, rather than by rational, self-conscious decision, as Cartesians would have it. Nevertheless, there is no reason to think that the genealogist would have to exempt the genealogist's self from its understanding of selfhood as such. The genealogical conception of the self changes along with changes in related concepts such as power and agency.

If MacIntyre thus criticizes Foucault for saying that individual subjectivity is completely constituted by external power relations, the rebuttal consists of showing that Foucault's writings shift in response from a genealogy of power to a genealogy of ethics, with its renewed interest in the self and subjectivity. There are at least two respects, however, in which this return to the concern for the first-person standpoint is not a return to classical phenomenology. First, phenomenology *starts* from consciousness, subjectivity, and agency. As Heidegger says in the 1925 Dilthey essay, phenomenology defines the human being as "a context of experiences held together through the unity of the ego as a center of acts."[40] The more recent interest in these phenomena in continental philosophy arises from seeing these first-person phenomena as constituted rather than as constituting, that is, as situated and embodied.[41] Ethics is where the self and freedom are issues, but in the hands of philosophers such as Foucault, Derrida, and Deleuze, the cogito, transcendental ego, or the "I think," is itself unmasked as false consciousness.

This discussion leads to the second respect in which genealogy is not classically phenomenological, even if Bernard Williams is right that it can include a phenomenological moment. Taking Foucault's philosophy as a paradigm, the concern with selfhood, subjectivity, and agency is not so much with how we identify with whomever we have been constructed to be. On the contrary, Foucault thinks that by showing us how we have been formed by external forces, we will resist and subvert the identities that we have been given. That is why, as we saw earlier, Foucault speaks of *désassujettissement*, which can be translated as either desubjec-tification or desubjugation.[42]

Genealogy's ability to unmask power relations is thus also an ability to desubjugate socially constituted subjects. Genealogy therefore has explicit transformative potential that the descriptive and reconstructive aims of phenomenology mute. If philosophy today intends not only to describe the way things are, but also to enable us to resist formations of the self that limit and distort our possibilities, then genealogy is an effective means for writing the kind of critical history that can lead to experimentation and self-transformation.[43]

As a closing self-reflection, I note that this book is itself more of a vindicatory exercise than an unmasking one. The project has been to examine everyday beliefs about temporality along with their philosophical interpretations, to deconstruct them or turn them in another direction, and to come up with a different analysis than would result from the standpoint of the metaphysics of universal time. So is the fact that this book vindicates much of what our various authors have written a shortcoming of the work? I think not. Vindication is sometimes exactly what is needed. Closer to Hume than to Nietzsche, vindication may not seem as profound or as earth shattering as successful unmasking. What this study attempts to exonerate, in any case, is the self-understanding of the genealogical method itself. Genealogy would not and should not resist such attempts to vindicate its usefulness, cogency, and coherence—in sum, its rationality.

Notes

Introduction

1. Ludwig Wittgenstein, *On Certainty* (Oxford: Blackwell, 1975), p. 141.

2. To give an idea of the historical chronology of the European philosophers mentioned in this book, here are the dates of their lives: Immanuel Kant (1724–1804), G. W. F. Hegel (1770–1831), Arthur Schopenhauer (1788–1860), Karl Marx (1818–1883), Wilhelm Dilthey (1833–1911), William James (1842–1910), Friedrich Nietzsche (1844–1900), Sigmund Freud (1856–1939), Edmund Husserl (1859–1938), Henri Bergson (1859–1941), Marcel Proust (1871–1922), Albert Einstein (1879–1955), Martin Heidegger (1889–1976), Walter Benjamin (1892–1940), Max Horkheimer (1895–1973), Hans-Georg Gadamer (1900–2002), Theodor Adorno (1903–1969), Jean-Paul Sartre (1905–1980), Emmanuel Lévinas (1906–1995), Simone de Beauvoir (1908–1986), Maurice Merleau-Ponty (1908–1961), Paul Ricoeur (1913–2005), Gilles Deleuze (1925–1995), Michel Foucault (1926–1984), Jürgen Habermas (b. 1929), Jacques Derrida (1930–2004), Pierre Bourdieu (1930–2002), Sarah Kofman (1934–1994), Julia Kristeva (b. 1941), Giorgio Agamben (b. 1942), Slavoj Žižek (b. 1949).

3. Thus, the principal continental authors studied in chapter 1 are Kant and Heidegger. Chapter 2 features Hegel, James, Husserl, Merleau-Ponty, Heidegger, Derrida, and Nietzsche, in that order. Chapter 3 includes a section on Husserl, Heidegger, and Gadamer, followed by another on Sartre, Bourdieu, and Foucault, as well as a third on Bergson as interpreted by Merleau-Ponty and Deleuze. Chapter 4 focuses on Kant, Hegel, Heidegger, Benjamin, Deleuze, Derrida, and Žižek. Chapter 5 summarizes the temporal strategies of Proust, Benjamin, Heidegger, Derrida, Žižek, and Deleuze.

1 In Search of Lost Time: Kant and Heidegger

1. Owen Flanagan advances this claim about Kant in *The Science of the Mind* (Cambridge, Mass.: MIT Press, 1984), pp. 180–184. See also Flanagan's *Consciousness Reconsidered* (Cambridge, Mass.: MIT Press, 1992).

2. See the "Transcendental Aesthetic" of Kant's *Critique of Pure Reason*, which is cited here in the translation by Paul Guyer and Allen W. Wood (Cambridge: Cambridge University Press, 1998), with occasional modifications of my own.

3. Kant, *Critique of Pure Reason*, Bxxxix. Heidegger comments in *Being and Time* that "the 'scandal of philosophy' is not that this proof has yet to be given, but that *such proofs are expected and attempted again and again*" (*Being and Time*, trans. John Macquarrie and Edward Robinson; New York: Harper and Row, 1962; p. 249; *Sein und Zeit,* p. 205).

4. Kant reaffirms in the Paralogisms of the *Critique of Pure Reason* that "we have in inner intuition nothing at all that persists, for the I is only the consciousness of my thinking" (B413). He again denies there that the awareness that "I exist thinking" has enough content to tell me anything about myself: "it is not possible at all through this simple self-consciousness to determine the way I exist, whether as substance or as accident" (B420).

5. On hallucination, see Maurice Merleau-Ponty's account in the *Phenomenology of Perception,* trans. Colin Smith (London: Routledge and Kegan Paul, 1962), pp. 334–345.

6. Martin Heidegger, *Kant and the Problem of Metaphysics*, 5th edition, trans. Richard Taft (Bloomington: Indiana University Press, 1997), p. xix.

7. Heidegger, *Being and Time*, p. 44 (SZ 22). (Throughout, German pagination will be cited as SZ).

8. Heidegger, *Kant*, p. xx.

9. Ibid.

10. Ibid., p. 119.

11. Ibid., p. 83.

12. Martin Heidegger, *The Metaphysical Foundations of Logic*, trans. Michael Heim (Bloomington: Indiana University Press, 1984), pp. 165–168. Intentionality and transcendence will be discussed in more detail in later chapters.

13. On the Husserl-Brentano polemic see Paul Ricoeur, *Time and Narrative,* volume 3, trans. Kathleen Blamey and David Pellauer (Chicago: University of Chicago Press, 1988), p. 30.

14. Heidegger, *Kant*, p. 122.

15. Ibid., p. 123.

16. Ibid.

17. Ibid., p. 131.

18. Ibid., p. 127.

19. Ibid.

20. Ibid., pp. 127–128.

21. Ibid., p. 130.

22. Ibid.

23. Cited by Heidegger, ibid., p. 129.

24. Ibid., p. 130.

25. Ibid.

26. Ibid., p. 132.

27. Ibid., pp. 132–133.

28. Ibid., p. 134.

29. Ibid.

30. Ibid.

31. Ibid.

32. Ibid.

33. Ibid., p. 141.

34. Ibid., p. 134.

35. Ibid., p. 140.

36. Ibid., p. 141.

37. Martin Heidegger, *Supplements: From the Earliest Essays to* Being and Time *and Beyond*, ed. John Van Buren (Albany: SUNY Press, 2003), p. 172.

38. Ibid., p. 172.

39. Ibid., p. 169.

40. Ibid.

41. For a reading of Heidegger as an idealist about time, see William D. Blattner, *Heidegger's Temporal Idealism* (Cambridge: Cambridge University Press, 1999).

42. Heidegger, *Being and Time*, p. 380 (SZ 331).

43. Robert Dostal also concludes from this passage that "Time is somehow prior to Dasein." See Robert J. Dostal, "Time and Phenomenology in Husserl and Heidegger," *The Cambridge Companion to Heidegger,* ed. Charles Guignon (Cambridge: Cambridge University Press, 2006), p. 165.

44. Heidegger, *Being and Time*, p. 377 (SZ 328).

45. Martin Heidegger, *The Basic Problems of Phenomenology*, trans. Albert Hofstadter (Bloomington: Indiana University Press, 1982), p. 325.

46. Ibid., p. 324.

47. Ibid., p. 236.

48. Ibid., p. 266.

49. Ibid., p. 324.

50. Ibid., p. 325.

51. For a very clear discussion of *The Fundamental Concepts of Metaphysics* and an explanation of the relevance of Heidegger's critique of ascertaining to current cognitive psychology, see Sue P. Stafford and Wanda Torres Gregory, "Heidegger's Phenomenology of Boredom, and the Scientific Investigation of Conscious Experience," *Phenomenology and the Cognitive Sciences* (2006) 5: 155–169.

52. Martin Heidegger, *The Fundamental Concepts of Metaphysics: World, Finitude, Solitude*, trans. William McNeill and Nicholas Walker (Bloomington: Indiana University Press, 1995), p. 65.

53. Ibid.

54. Ibid., p. 66.

55. Ibid., p. 88.

56. Ibid., p. 133.

57. Ibid., p. 138.

58. Ibid., p. 145.

59. Ibid., p. 174.

60. Ibid., p. 166.

61. Ibid., p. 172.

62. Ibid., p. 7.

63. Ibid.

64. Ibid., p. 141.

65. Martin Heidegger, "What Is Metaphysics?" in *Martin Heidegger: Basic Writings*, ed. David Farrell Krell (New York: Harper Collins, 1993), p. 99.

66. Heidegger, *Fundamental Concepts of Metaphysics,* p. 158.

67. Ibid., pp. 134–135.

68. See Thomas Nagel, "Death," in *Mortal Questions* (Cambridge: Cambridge University Press, 1979), pp. 1–10.

69. Heidegger, *Fundamental Concepts of Metaphysics,* p. 153. Emphasis removed.

70. Ibid., p. 144.

71. Ibid., p. 158.

72. Ibid.

73. The analytic philosophical tradition is divided between defenders of subjectivity such as Thomas Nagel, and critics of subjectivity such as Donald Davidson. Davidson in a series of papers on first-person access attacked what he labeled the "myth of subjectivity." Contemporary Husserlians often seem to be holding on to this myth, in contrast to Heideggerians who side more with Davidson.

74. "Cultural politics" is Richard Rorty's term for "arguments about what words to use." Richard Rorty, *Philosophy as Cultural Politics: Philosophical Papers* (Cambridge: Cambridge University Press, 2007), p. 3. In my terms, controversies in cultural politics are wars of words, if not of worlds.

2   There Is No Time Like the Present! On the Now

1. G. W. F. Hegel, *Phenomenology of Spirit,* trans. A. V. Miller (Oxford: Oxford University Press, 1977), p. 63.

2. Ibid., p. 64.

3. William James, *The Principles of Psychology*, volume 1 (New York: Dover, 1950), p. 606.

4. Ibid.

5. For a contemporary example of how to imagine such a life, consider the film *Memento.* In this film, the protagonist suffers brain damage and constantly loses his short-term memories. In an attempt to communicate with himself after each bout of amnesia wipes the slate of memories clean, the protagonist leaves himself notes and clues. However, he has a history of wildly misinterpreting them. He clearly does not know who he is, and his life is chaotic and unpredictable. Moreover, the film starts at the end and works its way back to the forgotten beginning, and the audience is left with the task of figuring out the "real" causal sequence.

6. James, *Principles of Psychology,* p. 607.

7. Ibid., pp. 608–609.

8. Ibid., p. 609.

9. Ibid., p. 610.

10. Ibid.

11. Ibid., p. 631. Emphasis removed.

12. Ibid., p. 630.

13. Ibid.

14. Ibid., p. 642.

15. Ernst Pöppel, quoted in "Connections," Edward Rothstein, *New York Times,* 10 January 2004, p. A15.

16. James, *Principles of Psychology,* p. 627.

17. Ibid., p. 638.

18. Ibid., p. 641.

19. Edmund Husserl, *On the Phenomenology of the Consciousness of Internal Time (1893–1917),* trans. John Barnett Brough (Dordrecht: Kluwer Academic Publishers, 1991).

20. According to John Brough, the scholarly consensus is that the period from late 1909 through 1911 is the more mature period of Husserl's theory of time-consciousness. In particular, §35 through §39 of the book that Heidegger edited and published for Husserl in 1928, along with supplementary text 54, appear to date from late 1911 and represent a good expression of his theory of duration. In Dan Zahavi's opinion in his 2003 book from the MIT Press, *Subjectivity and Selfhood: Investigating the First-Person Perspective,* where he discusses the *Bernauer Manuskripte über das Zeitbewusstsein* first published by Kluwer in 2001, "If there is anything that recent Husserl scholarship has demonstrated, it is that it is virtually impossible to acquire a satisfactory understanding of Husserl's views if one restricts oneself to the writings that were published during his lifetime" (p. 50). This attitude contrasts sharply with that of North American Nietzsche scholarship, where posthumous material is considered of dubious value unless it is supported by material published during Nietzsche's lifetime.

21. On Husserl's rejection of James's specious present, see Izchak Miller, *Husserl, Perception, and Temporal Awareness* (Cambridge, Mass.: MIT Press, 1984), pp. 170–171.

22. Husserl, *Internal Time*, p. 381. This text was incorporated into the published work edited by Heidegger with slight alterations. Compare p. 78.

23. Ibid., p. 390.

24. Husserl here seems to some critics to be caught between a punctual theory of experience and a durational theory. Although he does not subscribe to the view that time is a series of nows, he nevertheless often writes as if experience consisted of discrete moments of the Now.

25. In explaining Husserl's graph Izchak Miller suggests that we imagine the protentional line as "continuing a little way upwards and soon thereafter fading away." See Izchak Miller, *Husserl, Perception, and Temporal Awareness*, p. 137.

26. The details about protention can be found in *Die Bernauer Manuskripte über das Zeitbewusstsein (1917/18)* (Dordrecht: Kluwer, 2001).

27. Husserl, *Internal Time*, p. 390, n. 54.

28. The statement is by John Barnett Brough in his "Translator's Introduction" to Husserl's *Internal Time*, p. lii. For an earlier interpretation of these distinctions see Izchak Miller's *Husserl, Perception, and Temporal Awareness*.

29. Husserl, *Internal Time*, p. 60.

30. Ibid.

31. Ibid., pp. 84ff., 392.

32. Ibid., p. 392, n. 27.

33. Dan Zahavi, "Inner Time-Consciousness and Pre-reflective Self-awareness," in *The New Husserl: A Critical Reader*, ed. Donn Welton (Bloomington: Indiana University Press, 2003), p. 165.

34. Martin Heidegger, *Being and Time*, trans. John Macquarrie and Edward Robinson (New York: Harper and Row, 1962), p. 461 (SZ 408).

35. Ibid.

36. Ibid., p. 462 (SZ 409).

37. For a discussion of the importance, and the difficulties, of this distinction between two kinds of interpretation, see my essay "Post-Cartesian Interpretation: Hans-Georg Gadamer and Donald Davidson," in *The Philosophy of Hans-Georg Gadamer*, ed. Lewis E. Hahn (La Salle, Ill.: Open Court, 1997), pp. 111–128. See also my chapter, "Heidegger and the Hermeneutic Turn," in *The Cambridge Companion to Heidegger*, ed. Charles B. Guignon (Cambridge: Cambridge University Press, 1993), pp. 170–194.

38. Heidegger, *Being and Time*, p. 463 (SZ 410).

39. Ibid., p. 395 (SZ 345).

40. Ibid., p. 399 (SZ 348).

41. Ibid., p. 474 (SZ 422).

42. Ibid., p. 479 (SZ 426).

43. Ibid.

44. Ibid.

45. Ibid., p. 462 (SZ 409).

46. Ibid., p. 478 (SZ 426).

47. Ibid., p. 474 (SZ 422).

48. Ibid., p. 479 (SZ 427).

49. Ibid. Sartre also finds this metaphor of pregnancy useless and misleading. See *Being and Nothingness*, trans. Hazel E. Barnes (New York: Philosophical Library, 1956), p. 181.

50. Martin Heidegger, *The Metaphysical Foundations of Logic*, p. 165. On Heideggerian transcendence as a more basic condition than Husserlian intentionality, see pp. 167–168. In *Basic Problems of Phenomenology* Heidegger also confirms "that intentionality is founded in the Dasein's transcendence and is possible solely for this reason—that transcendence cannot conversely be explained in terms of intentionality" (p. 162).

51. Heidegger, *The Fundamental Concepts of Metaphysics*, p. 285.

52. Ibid., p. 286.

53. Maurice Merleau-Ponty, *Phenomenology of Perception*, p. 412.

54. Ibid.

55. Ibid., p. 428.

56. Ibid., p. 424.

57. Ibid.

58. Ibid., p. 421.

59. Ibid.

60. Ibid., p. 412.

61. Ibid., p. 414.

62. Ibid., p. 424.

63. Ibid., p. 430.

64. Ibid., p. 415.

65. See Stephen Priest, *Merleau-Ponty* (New York: Routledge, 1998), p. 131.

66. Merleau-Ponty, *Phenomenology of Perception*, p. 416.

67. Ibid., p. 417.

68. Ibid., p. 417.

69. Ibid., p. 418; emphasis added.

70. Ibid., p. 419.

71. Ibid., p. 423.

72. Ibid., p. 419.

73. Ibid., p. 421.

74. Alva Noë, in his book *Action in Perception* (Cambridge, Mass.: MIT Press, 2004), calls this the problem of the "presence of absence." He suggests that this

problem arises for the theorist of perception insofar as "we have a sense of the presence of that which, strictly speaking, we do not perceive" (p. 60).

75. Merleau-Ponty, *Phenomenology of Perception*, p. 417.

76. Ibid., p. 416.

77. Ibid.

78. Ibid., p. 411.

79. Ibid., pp. 419–420.

80. Ibid.

81. Ibid., p. 422.

82. Ibid., pp. 425–426 (emphasis added). Note the connection of "thrust" in this passage to the idea of "violent transition" on p. 382: "the violent transition from what I have to what I aim to have, from what I am to what I intend to be."

83. Maurice Merleau-Ponty, *The Visible and the Invisible, Followed by Working Notes*, trans. Alphonso Lingis (Evanston: Northwestern University Press, 1968), p. 190.

84. Merleau-Ponty, *Phenomenology of Perception*, p. 426.

85. Ibid., p. 428.

86. Ibid., p. 429.

87. Ibid., p. 430; citing Heidegger, *Sein und Zeit*, p. 366.

88. Merleau-Ponty, *Phenomenology of Perception*, p. 432; emphasis added.

89. Ibid., p. 433; emphasis added.

90. Ibid., pp. 420–421.

91. Ibid., p. 421.

92. Ibid., p. 424.

93. Ibid., p. 413.

94. Jacques Derrida, *Of Grammatology*, trans. Gayatri Chakravorty Spivak (Baltimore: The Johns Hopkins University Press, 1976), p. 65.

95. Ibid., p. 71.

96. *Derrida*, a film by Kirby Dick and Amy Ziering Kofman (Zeitgeist Video, 2002).

97. Jacques Derrida, *Speech and Phenomena, and Other Essays on Husserl's Theory of Signs*, trans. David B. Allison (Evanston: Northwestern University Press, 1973), pp. 79–80. (*La Voix et le phénomène* [Paris: Presses Universitaires de France, 1967], p. 89.)

98. Ibid., p. 79. (*La Voix*, p. 89.)

99. Ibid., p. 85. (*La Voix*, p. 95.)

100. Derrida, *Of Grammatology*, p. 68.

101. Ibid., p. 69.

102. Ibid., p. 65–66 (emphasis added).

103. Ibid., p. 65.

104. Ibid., p. 69.

105. Ibid., p. 65.

106. Ibid., p. 67.

107. Ibid.

108. Ibid., p. 72. For a defense of Husserl against that charge of being guilty of the metaphysics of presence, see Dan Zahavi, *Husserl's Phenomenology* (Stanford: Stanford University Press, 2003), pp. 93–98. See also Dan Zahavi, *Self-Awareness and Alterity: A Phenomenological Investigation* (Evanston, Ill.: Northwestern University Press, 1999), pp. 82–87.

109. Friedrich Nietzsche, *The Gay Science*, trans. Josefine Nauckhoff (Cambridge: Cambridge University Press, 2001), p. 194.

110. Friedrich Nietzsche, *The Will to Power*, trans. Walter Kaufmann and R. J. Hollingdale (New York: Vintage, 1968), p. 341. (See Schacht, *Nietzsche, Genealogy, Morality: Essays on Nietzsche's "Genealogy of Morals"* [Berkeley: University of California Press, 1994], pp. 253ff.)

111. Gilles Deleuze, *Pure Immanence: Essays on a Life*, trans. Anne Boyman (New York: Zone Books, 2001), p. 87.

112. Ibid., p. 88.

113. Elizabeth Grosz, *The Nick of Time: Politics, Evolution, and the Untimely* (Durham, N.C.: Duke University Press, 2004), p. 142.

114. Ibid., p. 143.

115. Nietzsche, *Gay Science*, p. 110.

116. Gilles Deleuze, *Nietzsche and Philosophy*, trans. Hugh Tomlinson (London: Athlone Press, 1983), p. 68.

117. *Thus Spoke Zarathustra* 2: 139. Quoted from *The Nietzsche Reader*, ed. Keith Ansell Pearson and Duncan Large (Oxford: Blackwell, 2006), p. 275.

118. Friedrich Nietzsche, *Ecce Homo* ("Why I Am So Clever," §10). Quoted from *The Nietzsche Reader*, p. 509.

119. See Grosz, *Nick of Time*, p. 151.

120. "Others" includes but is not limited to "other humans."

121. See Sartre, "The Wall," in *The Wall (Intimacy) and Other Stories,* trans. Lloyd Alexander (New York: New Directions, 1975).

122. Alexander Nehamas, *Nietzsche: Life as Literature* (Cambridge, Mass.: Harvard University Press, 1985), p. 190.

123. Nietzsche, *Zarathustra*, 4, 19. Cited by Nehamas, *Nietzsche*, p. 190.

124. Heidegger, *Being and Time*, p. 382 (SZ 333).

125. The example is from Gilles Deleuze, *Difference and Repetition*, trans. Paul Patton (New York: Columbia University Press, 1994), p. 72.

3   Where Does the Time Go? On the Past

1. Michael Dummett tells us that there are four possible metaphysical stances on the reality of the past: (1) only the present is real, and the past and the future are not real; (2) the past and the future are real, as is the present; (3) the past is real, as is the present, but the future is not real; (4) the future is real, as is the present, but the past is not real. Michael Dummett, *Truth and the Past* (New York: Columbia University Press, 2004). We will see in chapter 5 an additional metaphysics of time, that of Gilles Deleuze. Deleuze would say that the important contrast is between the first of these four stances (he calls it "Chronos") and a fifth stance that says that the past and the future are what are real, and the present is not real, but only virtual (he calls it "Aion"). See Gilles Deleuze, *The Logic of Sense*, trans. Mark Lester (New York: Columbia University Press, 1990), pp. 162–168.

2. Hayden White, "The Burden of History," *History and Theory* (1966) 5: 111–134, p. 123. Reprinted in *Tropics of Discourses* (Johns Hopkins University Press, 1978), p. 39.

3. Paul Ricoeur quotes Ranke in volume 3 of *Time and Narrative*, trans. Kathleen Blamey and David Pellauer (Chicago: University of Chicago Press, 1988), p. 185.

4. See my forthcoming essay, "One What? Relativism and Poststructuralism," in *Relativism: A Compendium*, ed. Michael Krausz (New York: Columbia University Press, 2008).

5. Ian Hacking, *Rewriting the Soul: Multiple Personality and the Sciences of Memory* (Princeton: Princeton University Press, 1995), p. 243.

6. Ian Hacking, "Language, Truth, and Reason," in *Historical Ontology* (Cambridge, Mass.: Harvard University Press, 2002), p. 160.

7. Michel Foucault, "What Is an Author?" in *The Foucault Reader*, ed. Paul Rabinow (New York: Pantheon, 1984), p. 114.

8. Hacking, *Historical Ontology*, p. 175.

9. Heidegger speaks of "factical possibilities" at German pages 269, 299, and especially page 383 of *Sein und Zeit* (see *Being and Time*, p. 434).

10. James, *Principles of Psychology*, p. 627.

11. Perhaps at this early stage of the analysis it will be sufficient to say that the memory involves an awareness that the content is not directly perceived, and that the content is dated. The *datability* of content keeps us from confusing the experience of an earlier content with the experience of a later content.

12. Husserl, no. 53: "The Intentionality of Internal Consciousness," *Internal Time*, pp. 370–379.

13. Husserl, *Internal Time*, p. 378.

14. Ibid., p. 370. See pp. 29–30.

15. Ibid., p. 88.

16. Ibid.

17. Ibid., p. 56.

18. Ibid.

19. Ibid., p. 67.

20. See Martin Heidegger, *Supplements*, p. 172.

21. Heidegger, *Being and Time*, p. 373 (SZ 326).

22. I note that the phrase "to deal with it" is ambiguous between inner and outer dealing. In ordinary usage it means to do something objective and concrete about the situation (e.g., if you have debts, pay them off). However, in current slang, where it contrasts to being "in denial" about some aspect of the situation, it just means being aware of (as opposed to being blind to) the problem, but not necessarily doing anything about it. That is, it has more to do with inner attitude than outer action.

23. By "facts" I mean what Heidegger calls "factuality," i.e., facts about things, as opposed to "facticity," which involves human possibilities.

24. For insightful studies of the foibles of memory, see Ian Hacking, *Rewriting the Soul*.

25. Jacques Derrida, *Without Alibi*, trans. Peggy Kamuf (Stanford: Stanford University Press, 2002), p. 169.

26. Ibid., p. 163.

27. Jean-Paul Sartre, *Being and Nothingness*, p. 639.

28. Ibid., p. 638.

29. Ibid.

30. Ibid., p. 640.

31. Ibid., p. 641.

32. Ibid., p. 642.

33. Ibid., p. 643.

34. Ibid., p. 645.

35. Ibid., emphasis added.

36. Ibid., p. 280.

37. Ibid., p. 285.

38. Ibid.

39. Ibid., p. 646.

40. Ibid., p. 784.

41. Pierre Bourdieu, *The Logic of Practice*, trans. Richard Nice (Stanford: Stanford University Press), p. 62.

42. Pierre Bourdieu, *Pascalian Meditations*, trans. Richard Nice (Stanford: Stanford University Press), p. 206.

43. Ibid., p. 208.

44. Ibid., p. 213.

45. Maurice Merleau-Ponty, *In Praise of Philosophy*, trans. John Wild and James M. Edie (Evanston, Ill.: Northwestern University Press, 1963). See also Merleau-Ponty, *Nature: Course Notes from the Collège de France*, trans. Robert Vallier (Evanston, Ill.: Northwestern University Press, 2003) and Merleau-Ponty, *The*

Incarnate Subject: Malebranche, Biran, and Bergson on the Union of Body and Soul, trans. Paul B. Milan (New York: Humanity Books, 2001). For a summary of the confluence of the ideas of Merleau-Ponty and Bergson, see Mark S. Muldoon, *Tricks of Time: Bergson, Merleau-Ponty and Ricoeur in Search of Time, Self, and Meaning* (Pittsburgh, Penn.: Duquesne University Press, 2006), pp. 119–122. For valuable accounts of Bergson and Deleuze on time, see Elizabeth Grosz, *The Nick of Time: Politics, Evolution, and the Untimely* (Durham, N.C.: Duke University Press, 2004), pp. 155–243; Giovanna Borradori, "The Temporalization of Difference: Reflections on Deleuze's Interpretation of Bergson" (*Continental Philosophy Review* 34: 1–20, 2001); Constantin V. Boundas, "Deleuze—Bergson: an Ontology of the Virtual," in *Deleuze: A Critical Reader*, ed. Paul Patton (Oxford: Blackwell, 1996); Suzanne Guerlac, *Thinking in Time: An Introduction to Henri Bergson* (Ithaca, N.Y.: Cornell University Press, 2006); Keith Ansell-Pearson, *Philosophy and the Adventure of the Virtual: Bergson and the Time of Life* (London: Routledge, 2002); James Williams, *Gilles Deleuze's* Difference and Repetition: *A Critical Introduction and Guide* (Edinburgh: Edinburgh University Press, 2003); Brian Massumi, *Parables for the Virtual: Movement, Affect, Sensation* (Durham, N.C.: Duke University Press, 2002); John Mullarky, *Bergson and Philosophy* (Edinburgh: Edinurgh University Press, 1999) as well as *The New Bergson* (Manchester: Manchester University Press, 1999).

46. Merleau-Ponty, *In Praise of Philosophy*, p. 21.

47. Ibid.

48. Merleau-Ponty, *Nature*, p. 70.

49. Merleau-Ponty, *In Praise of Philosophy*, p. 12.

50. Ibid., pp. 14–15 (translation amended).

51. Merleau-Ponty's metaphor of "haunting" and "ghosts" may well echo Bergson's own penchant for this type of figure of speech. For instance, in *Matière et mémoire* Bergson speaks of a memory as a *revenant* or ghostlike apparition: "Alors, quand un souvenir reparaît à la conscience, il nous fait l'effet d'un revenant dont il faudrait expliquer par des causes spéciales l'apparition mystérieuse" (p. 161).

52. Merleau-Ponty, *Phenomenology of Perception*, p. 79 (n. 2).

53. Ibid.

54. Ibid., p. 415.

55. Ibid., p. 276.

56. Ibid., p. 420.

57. Bergson, *Durée et simultanéité*, p. 179; cited by Merleau-Ponty, *Nature*, p. 111.

58. Merleau-Ponty, *Nature*, pp. 108–109.

59. Ibid., p. 109.

60. Ibid., p. 112.

61. Ibid., p.111.

62. Ibid., p. 112.

63. Merleau-Ponty, *In Praise of Philosophy*, p. 22.

64. Cited by Merleau-Ponty, *In Praise of Philosophy*, p. 16.

65. Ibid., p. 17.

66. Henri Bergson, *Matière et mémoire* (Paris: PUF, 1939), p. 168.

67. Deleuze, *Bergsonism*, p. 54. Emphasis removed.

68. Henri Bergson, *Matter and Memory*, trans. N. M. Paul and W. S. Palmer (New York: Zone Books, 1991), pp. 148–149.

69. Deleuze, *Bergsonism*, p. 55.

70. Ibid., p. 58.

71. Bergson, *L'Energie spirituelle*, cited by Deleuze, pp. 125–126.

72. Bergson, *Matière et mémoire*, p. 167 (my translation).

73. Gilles Deleuze, *Difference and Repetition*, trans. Paul Patton (New York: Columbia University Press, 1994), pp. 81–82.

74. Ibid., p. 82.

75. Ibid.

76. Ibid.

77. Earlier, however, Deleuze had said that "it is memory that grounds time" (*Difference and Repetition*, p. 79). This seems to imply what I consider an even greater paradox, namely, that what we remember is the past that was never present. Of course, if memory is only ever partial, then the claim is less surprising insofar as it says merely that the past is always more complex than it can be represented to be.

78. Deleuze, *Bergsonism*, pp. 61–62.

79. Ibid., p. 57.

80. Ibid., p. 59.

81. Ibid., p. 71.

82. Deleuze, *Difference and Repetition*, p. 115.

83. Sarah Kofman, the French feminist philosopher and Nietzsche scholar, develops the implications of the poststructuralist reading of Nietzsche as a pluralist in *Nietzsche and Metaphor*, trans. Duncan Large (Stanford: Stanford University Press, 1993). See my discussion in chapter 1 of *Critical Resistance: From Poststructuralism to Post-Critique* (Cambridge, Mass.: MIT Press, 2004).

84. Deleuze, *Bergsonism*, pp. 81–82.

85. Ibid., p. 82.

86. Ibid., pp. 104–105.

87. Heidegger, *Basic Problems of Phenomenology*, p. 328.

4 "The Times They Are a-Changin' ": On the Future

1. Immanuel Kant, *On History*, ed. Lewis White Beck (Indianapolis: Bobbs-Merrill, 1963), p. 112 (emphasis added).

2. Ibid., p. 9.

3. Ibid.

4. Ibid., p. 4.

5. Kant here has a specific freedom in mind: freedom of expression.

6. G. W. F. Hegel, *Lectures on the Philosophy of World History: Introduction*, trans. H. B. Nisbet (Cambridge: Cambridge University Press, 1975), p. 77.

7. Heidegger, *Being and Time*, p. 395 (SZ 344).

8. Ibid., p. 394 (SZ 344).

9. Ibid., p. 397 (SZ 347).

10. The retro is not identical to nostalgia. Young people adopt a retro look of the late '60s, for example, without any awareness of what it was like to have lived then. To the young people the retro is simply "a look." Of course, to say that they do not know what it was really like then is nostalgic now.

11. Heidegger, *Being and Time*, p. 401 (SZ 350).

12. Ibid., p. 373 (SZ 326).

13. Ibid.

14. Ibid., p. 388 (SZ 338).

15. Walter Benjamin, *Illuminations*, trans. Harry Zohn (New York: Schocken, 1969), pp. 257–258.

16. Ibid., p. 256.

17. Ibid., p. 254.

18. Ibid., p. 263.

19. Ibid., p. 260.

20. Ibid., p. 257.

21. Ibid., p. 255 (emphasis added).

22. Ibid., p. 260.

23. Ibid., pp. 257, 262–263.

24. Ibid., pp. 261–262.

25. Deleuze, *Difference and Repetition*, p. 75 (emphasis added).

26. Ibid., p. 78.

27. Ibid., p. 79.

28. Ibid.

29. Ibid., p. 78.

30. Keith W. Faulkner discovers the idea of the "larval ego" in Freud's *Project for a Scientific Psychology* and gives it a more psychoanalytic reading in his book, *Deleuze and the Three Syntheses of Time* (New York: Peter Lang, 2006), pp. ix, 65.

31. Deleuze, *Difference and Repetition*, p. 90.

32. Ibid., p. 91.

33. Ibid., p. 86.

34. Ibid., p. 87.

35. For a detailed analysis of Deleuze's "reinvention" of Kant's three syntheses, see Keith W. Faulkner's *Deleuze and the Three Syntheses of Time*, pp. 14ff.

36. Deleuze, *Difference and Repetition*, p. 94.

37. Ibid., p. 79.

38. Ibid., p. 94.

39. Ibid., p. 115.

40. Ibid.

41. Ibid., p. 126.

42. Ibid.

43. Ibid., p. 105.

44. Derrida, "Force of Law: The 'Mystical Foundation of Authority,'" in *Deconstruction and the Possibility of Justice*, ed. Drucilla Cornell, Michel Rosenfeld, and David Grey Carlson (New York: Routledge, 1992), p. 62.

45. See chapter 4 of my book, *Critical Resistance.*

46. Jacques Derrida, "Marx and Sons," in *Ghostly Demarcations: A Symposium on Jacques Derrida's* Specters of Marx, ed. Michael Sprinker (London and New York: Verso, 1999), p. 242.

47. Jacques Derrida, "The Three Ages of Jacques Derrida: An Interview with the father of Deconstructionism," with Kristine McKenna, *LA Weekly*, November 8–14, 2002 (available online at www.laweekly.com).

48. Jacques Derrida, *Rogues: Two Essays on Reason*, trans. Pascale-Anne Brault and Michael Naas (Stanford: Stanford University Press, 2005), p. 110.

49. I owe this term to Hubert Dreyfus.

50. Heidegger, *Der Spiegel Interview*; cited by Derrida, *Rogues*, p. 111.

51. Heidegger, *Der Spiegel Interview*; cited by Derrida, *Rogues*, p. 112.

52. Derrida, *Rogues*, p. 82.

53. Ibid., p. 83.

54. Ibid., p. 84.

55. Ibid., pp. 84–85.

56. See my discussion of Onora O'Neill and Christine Korsgaard in "The Ethics of Freedom: Hegel on Reason as Law-Giving and Law-Testing," in *Blackwell Guide to Hegel's* Phenomenology of Spirit, ed. Kenneth Westphal (Oxford: Blackwell, 2008).

57. Derrida, *Rogues*, p. 85.

58. Jacques Derrida, *The Other Heading*, trans. Pascale-Anne Brault and Michael Naas (Bloomington: Indiana University Press, 1992), p. 78.

59. Derrida, *Rogues*, p. 86.

60. Ibid.

61. Ibid.

62. Ibid., p. 87.

63. Ibid.

64. Ibid.

65. On p. 89 of *Rogues* Derrida cites himself from *Politics of Friendship*, pp. 103–104.

66. Derrida, *Rogues*, p. 90, citing *Politics of Friendship*, p. 105.

67. Derrida, *Rogues*, p. 90.

68. Ibid., p. 108.

69. Ibid., p. 109; emphasis added.

70. The term is Peter Sloterdijk's.

71. Jacques Derrida, *The Gift of Death*, trans. David Wills (Chicago: University of Chicago Press, 1995), pp. 74–75; emphasis added.

72. Gilles Deleuze, "Bartleby; or, The Formula" in *Essays Critical and Clinical*, translated by Michael A. Greco (Minneapolis: University of Minnesota Press, 1997), p. 71.

73. Ibid., front matter.

74. Ibid., p. 70.

75. Ibid., p. 71.

76. Ibid., p. 74.

77. Ibid., p. 73.

78. Ibid., p. 83.

79. Žižek, *The Parallax View* (Cambridge, Mass.: MIT Press, 2006), p. 347.

80. Ibid., p. 246.

81. Ibid., p. 282.

82. Ibid., p. 383.

83. For Žižek in *The Parallax View* (p. 284), there are three ways to read Heidegger's infamous statement about the "inner greatness" of National Socialism as the encounter between modern man and technology: (1) The Nazi project is an authentic act of assuming our destiny of challenging technology and undermining its nihilism; (2) Nazi total mobilization is more appropriate than liberal democracy to the essence of technology; (3) Nazism is modern nihilism at its most destructive and demoniac. These correspond to Heidegger's disillusionment with Nazism but also with the fate of the German war effort.

84. Ibid., p. 284.

85. Ibid., p. 378.

86. What do you do with human rights when they are of no use any longer? Like old clothes, quips Jacques Rancière, you give them to the poor, to people who do not have the power to use them. Cited as a "very striking dialectical reversal" in *Parallax View*, p. 341.

87. Žižek, *Parallax View*, p. 378.

88. Ibid., p. 342.

89. Ibid., p. 382.

90. Jacques Derrida and Maurizio Ferraris, *A Taste for the Secret*, trans. Giacomo Donis (Cambridge: Polity Press, 2001), p. 84.

91. See Derrida, "Marx and Sons," *Ghostly Demarcations*, p. 229.

5 *Le temps retrouvé*: Time Reconciled

1. The inspiration for the term "reconciliation" comes from Julia Kristeva's book, *Proust and the Sense of Time*, trans. Stephen Bann (New York: Columbia University Press, 1993). She identifies "psychic time" as the "space of reconciliation" (p. 4).

2. Michel Foucault, *Discipline and Punish: The Birth of the Prison*, trans. A. Sheridan (New York: Pantheon, 1977), pp. 30–31.

3. See my discussion of Habermas with Thomas McCarthy in our book, *Critical Theory* (London and New York: Blackwell, 1994).

4. Merleau-Ponty, *Phenomenology of Perception*, p. 422.

5. Arthur Schopenhauer, "The Vanity of Existence." Quoted from an online translation at http:Adelaide.edu.au/s/Schopenhauer/Arthur/. See Arthur Schopenhauer, *Essays and Aphorisms*, trans. R. J. Hollingdale (London: Penguin, 1970), pp. 51–54.

6. Ibid.

7. Ibid.

8. Deleuze, *Difference and Repetition*, p. 85; emphasis added; see p. 122.

9. Deleuze, *Proust and Signs*, trans. Richard Howard (Minneapolis: University of Minnesota Press, 2000), p. 56. In Deleuze's own words, "Joie si puissante qu'elle suffit à nous rendre la mort indifférente." Deleuze, *Proust et les signes* (Paris: PUF, 1964), p. 71.

10. Deleuze, *Proust*, p. 58 (French, p. 73).

11. Ibid., pp. 60–61 (French, p. 76).

12. In Deleuze's words, primordial temporality abruptly breaks into (*survient brusquement*) time that is already deployed. Ibid., p. 62 (French, p. 78).

13. Although it is not especially relevant, note that what set off Proust's reminiscence was not the madeleine itself, but rather the taste of the spoonful of tea into which he had dipped the madeleine. Note also that the family of Proust had a Paris domicile near the Place de la Madeleine.

14. Alain Badiou, *Deleuze: "La clameur de l'Être"* (Paris: Hachette, 1997), pp. 92–93.

15. Ibid., p. 93.

16. See Alexander Nehamas, *Nietzsche: Life as Literature* (Cambridge, Mass.: Harvard University Press, 1985).

17. Giorgio Agamben, *Infancy and History: On the Destruction of Experience*, trans. Liz Heron (New York: Verso, 2007), p. 99.

18. Ibid., p. 112.

19. Ibid., p. 115.

20. Ibid.

21. Ibid., p. 113.

22. For a discussion of historicity see David Couzens Hoy, "History, Historicity, and Historiography in *Being and Time*," in *Heidegger and Modern Philosophy: Critical Essays*, ed. Michael Murray (New Haven: Yale University Press, 1978), pp. 329–353.

23. Heidegger, *Being and Time*, p. 76 (SZ 50).

24. Ibid., p. 355 (SZ 380).

25. Heidegger, "Letter on Humanism," in *Martin Heidegger: Basic Writings*, p. 238.

26. Ibid., p. 240.

27. Ibid., p. 223; translation modified.

28. Ibid., pp. 234, 245.

29. The contrast here is noted by Vincent Descombes in *Modern French Philosophy* (Cambridge: Cambridge University Press, 1980), p. 30.

30. See Michel Foucault, "On the Genealogy of Ethics: An Overview of Work in Progress" as well as "The Ethics of the Concern of the Self as a Practice of Freedom" in *The Essential Foucault: Selections from Essential Works of Foucault 1954–1984*, ed. Paul Rabinow and Nikolas Rose (New York: New Press, 2003), pp. 102–125 and 18–24.

31. Grosz, *Nick of Time*, p. 257.

32. Wilhelm Reich was a psychoanalyst who lived from 1897 to 1957. Foucault mentions him on pages 5 and 131 of *The History of Sexuality*, volume 1: *An Introduction*, translated by Robert Hurley (New York: Pantheon, 1978).

33. Michel Foucault, *Ethics: Subjectivity and Truth*, ed. Paul Rabinow (The New Press, 1997), p. 298.

34. Ibid.

35. See Michel Foucault, *Psychiatric Power: Lectures at the Collège de France, 1973–1974*, trans. Graham Burchell (New York: Palgrave Macmillan, 2006), p. 56.

36. Ibid., p. 57.

37. Foucault, *Ethics: Subjectivity and Truth*, p. 284; translation modified.

38. Ibid.

39. Michel Foucault, *Psychiatric Power*, p. 47.

40. Ibid., p. 55.

41. Michel Foucault, "Structuralism and Post-Structuralism," *The Essential Foucault: Selections from the Essential Works of Foucault: 1954–1984*, ed. Paul Rabinow and Nikolas Rose (New York: The New Press, 1994), p. 93.

42. Ibid., p. 94.

43. Ibid.

44. Jacques Derrida, *Rogues*, p. 39.

45. Slavoj Žižek, *The Ticklish Subject: The Absent Centre of Political Ontology* (London: Verso, 1999), p. 23.

46. Slavoj Žižek, "Holding the Place," in Judith Butler, Ernesto Laclau, Slavoj Žižek, *Contingency, Hegemony, Universality: Contemporary Dialogues on the Left* (London: Verso, 2000), p. 321.

47. Slavoj Žižek, *The Sublime Object of Ideology* (London: Verso, 1989), p. 49.

48. Ibid., p. 36.

49. Ibid., p. 45.

50. Gilles Deleuze, *The Logic of Sense*, trans. Mark Lester (New York: Columbia University Press, 1990), p. 5.

51. Ibid.

52. Ibid.

53. Ibid.

54. Ibid., p. 166.

55. Ibid., p. 168.

56. Leonard Lawlor sees Chronos as Husserlian in contrast to my interpretation of Chronos as Bergsonian. I think, though, that my reading of Chronos as Bergsonian captures more of the text and is thus plausible. See Leonard Lawlor, "The Beginnings of Thought: The Fundamental Experience in Derrida and Deleuze," in *Between Deleuze and Derrida*, ed. Paul Patton and John Protevi (New York: Continuum, 2003), p. 76. I note that Keith Ansell Pearson sees Nietzsche and Kant as inhabitants of the time of Aion. See Keith Ansell Pearson, *Philosophy and the Adventure of the Virtual: Bergson and the Time of Life* (New York: Routledge, 2002), p. 197. Tamsin Lorraine also associates Nietzsche with the time of Aion in her essay, "Living a Time Out of Joint" (in *Between Deleuze and Derrida*, ed. Patton and Protevi; see p. 32).

57. Rudolf Bernet, "Husserl," in *A Companion to Continental Philosophy*, ed. Simon Critchley and William R. Schroeder (Oxford: Blackwell, 1998), pp. 203–204.

58. Deleuze, *The Logic of Sense*, p. 5.

59. Gilles Deleuze, *Cinema 1: The Movement-Image*, trans. Hugh Tomlinson and Barbara Habberjam (Minneapolis: University of Minnesota Press, 1986), pp. 60–61. In *The Imagination* (trans. F. Williams, Ann Arbor: University of Michigan Press, 1962, p. 40) Sartre had written, "We have here a sort of reversal of the classical comparison: instead of consciousness being a light going from the subject to the thing, it is a luminosity which goes from the thing to the subject" (cited by Deleuze, *Cinema I*, p. 227).

60. Gilles Deleuze, *Cinema 2: The Time Image*, trans. Hugh Tomlinson and Robert Galeta (Minneapolis: University of Minnesota Press, 1989), p. 81. I am adding the emphasis on "non-*chrono*logical" to bring out the unheard difference between Cronos and Chronos that Deleuze is playing on here.

61. Ibid.

62. Ibid., p. 82.

63. Ibid.

64. Ibid.

65. Ibid.

66. Deleuze, *The Logic of Sense*, p. 5.

67. Ibid., p. 61.

68. Ibid., p. 5. In *Mille plateaux* Deleuze and his coauthor, Félix Guattari, distinguish between Aeon and Chronos as longitude and latitude. They wish to avoid, however, any suggestion of an "oversimplified conciliation" between these "two modes of temporality." See G. Deleuze and F. Guattari, *A Thousand Plateaus: Capitalism and Schizophrenia*, trans. Brian Massumi (Minneapolis: University of Minnesota Press, 1987), pp. 262–263.

69. Ibid., p. 60.

70. In his 1978 book, *Infanzia e storia*, Giorgio Agamben draws on different ancient sources of the Chronos–Aion relation, but comes to conclusions that resemble Deleuze's. Agamben discovers the representation of the Chronos–Aion relation in Heraclitus and Plato. Agamben points to Plato's *Timaeus* where the relationship between Chronos and Aion is the relationship of copy and model as well as of cyclical, diachronic time and motionless, synchronic time. The point is not so much that "*aion* should be identified with eternity and *chrónos* with diachronic time as that our culture should conceive from its very origins a split between two different, correlated, and opposed notions of time." Agamben finds it significant that even earlier, at the very origins of European thought, a fragment of Heraclitus figures *Aion*, "time in its original sense," as a child playing with dice. Time is reconciled with temporality insofar as it is also perceived in the fragment as "the *temporalizing essence* of the living being." With this interpretation of Heraclitus, Agamben strikes me as reconciling the opposition between these two conflicting conceptions of *time* much as Deleuze does, namely, by turning them into conceptions of *temporality*. See Giorgio Agamben, *Infancy and History*, pp. 81–82.

71. Friedrich Nietzsche, *The Will to Power*, trans. Walter Kaufmann and R. J. Hollingdale (New York: Random House, 1967), p. 549.

72. Gilles Deleuze, *Nietzsche and Philosophy*, trans. Hugh Tomlinson (London: Athlone Press, 1983), p. 26.

Postscript on Method: Genealogy, Phenomenology, Critical Theory

1. See David Couzens Hoy, "Nietzsche, Hume, and the Genealogical Method," in *Nietzsche, Genealogy, Morality: Essays on Nietzsche's "Genealogy of*

Morals," ed. Richard Schacht (Berkeley: University of California Press, 1994), pp. 251–268.

2. Bernard Williams, *Truth and Truthfulness* (Princeton: Princeton University Press, 2002), p. 36.

3. Merleau-Ponty, *In Praise of Philosophy*, pp. 9–33.

4. Maurice Merleau-Ponty, *The Phenomenology of Perception*, p. vii.

5. Ibid.

6. Ibid., p. xx.

7. Richard Rorty, "A Pragmatist View of Contemporary Analytic Philosophy," in *The Pragmatic Turn in Philosophy: Contemporary Engagements between Analytic and Continental Thought*, ed. William Egginton and Mike Sandbothe (Albany: SUNY Press, 2004), pp. 131–144; quotation from p. 141.

8. Max Horkheimer, "Traditional and Critical Theory," in *Critical Theory: Selected Essays*, trans. Matthew J. O'Connell (New York: Continuum, 1992).

9. Ibid., p. 207.

10. Ibid., p. 199.

11. Ibid., p. 218.

12. Michel Foucault, "Polemics, Politics, and Problematizations: An Interview," in *The Foucault Reader*, ed. Paul Rabinow (New York: Pantheon, 1984), p. 384.

13. Horkheimer, "Traditional and Critical Theory," p. 227n.

14. Martin Jay, *Marxism and Totality: The Adventures of a Concept from Lukács to Habermas* (Berkeley: University of California Press, 1984), p. 210.

15. Horkheimer, "Traditional and Critical Theory," p. 200.

16. Ibid., p. 241.

17. Jacques Derrida, "Marx and Sons," p. 249. On messianicity see my discussion in *Critical Resistance,* pp. 186–190.

18. Clifford Geertz, "Stir Crazy," *New York Review of Books*, 26 January 1978.

19. Horkheimer, "Traditional and Critical Theory," p. 232.

20. Ibid., p. 242.

21. See chapter 5 of Hoy, *Critical Resistance.*

22. Alasdair MacIntyre, *Three Rival Versions of Moral Enquiry: Encyclopedia, Genealogy, and Tradition* (Notre Dame: University of Notre Dame Press, 1990), p. 46.

23. Jacques Derrida, *Aporias* (Stanford: Stanford University Press, 1993), p. 18; *The Other Heading: Reflections on Today's Europe* (Bloomington: Indiana University Press, 1992), p. 77. On deconstructive genealogy see Hoy, *Critical Resistance*, pp. 227–239.

24. Michel Foucault, "Preface to *The History of Sexuality, Volume Two*," in *The Foucault Reader*, ed. Paul Rabinow (New York: Pantheon, 1984), p. 335.

25. Ibid.

26. Michel Foucault, "Useless to Revolt?" in *Power: Essential Works of Foucault: 1954–1984*, ed. James D. Faubion, trans. Robert Hurley (New York: The New Press, 2000), p. 453 (emphasis added). For further discussion of Foucault and temporality, see my forthcoming essay, "The Temporality of Power," in *Engagement and Its Lack: Assessing Foucault's Legacy*, ed. Carlos Prado (New York: Continuum Press, 2009.)

27. Michel Foucault, *The Birth of Biopolitics: Lectures at the Collège de France, 1978–79*, ed. Michel Senellart (New York: Palgrave Macmillan, 2008), p. 3.

28. Michel Foucault, "Foucault by Maurice Florence," in *Michel Foucault: Aesthetics, Method, and Epistemology*, ed. James D. Faubion (New York: The New Press, 1998), p. 461.

29. Michel Foucault, *Security, Territory, Population: Lectures at the Collège de France, 1977–78*, trans. Graham Burchell (New York: Palgrave Macmillan, 2007), p. 118.

30. Foucault, *Discipline and Punish*, p. 202.

31. Derrida, *Aporias*, p. 19; *The Other Heading*, p. 78.

32. Derrida, *The Other Heading*, p. 73.

33. I draw on Pierre Bourdieu, who in *The Rules of Art: Genesis and Structure of the Literary Field* (Stanford: Stanford University Press, 1996) sees intellectuals as always aspiring to universals, but at the same time as always ventriloquizing the "historical unconscious" of a "singular intellectual field" (p. 340). See note 25.

34. This characterization of phenomenology comes from Sean Kelly of Harvard University in an unpublished paper presented at the 2002 NEH Summer Institute on "Consciousness and Intentionality" in Santa Cruz.

35. Bernard Williams, "Naturalism and Genealogy," in *Morality, Reflection, and Ideology*, ed. Edward Harcourt (Oxford: Oxford University Press, 2000), pp. 148–161; see p. 158.

36. MacIntyre, *Three Rival Versions*, p. 40.

37. Williams, "Naturalism and Genealogy," p. 158.

38. MacIntyre, *Three Rival Versions*, p. 55.

39. Ibid., p. 54.

40. Martin Heidegger, "Wilhelm Dilthey's Research and the Struggle for a Historical Worldview," in *Supplements: From the Earliest Essays to* Being and Time *and Beyond*, ed. John van Buren (Albany: SUNY Press, 2002), p. 162.

41. Witness the strong tradition of feminist critical phenomenology, with its focus on embodiment, beginning in 1949 with the publication in French of Simone de Beauvoir's *The Second Sex* (trans. H. M. Parshley, New York: Vintage Books, 1952) and including such American phenomenologists as Iris Young (*Throwing Like a Girl and Other Essays*, Bloomington: Indiana University Press, 1990), Sandra Bartky (*Femininity and Oppression: Studies in the Phenomenology of Oppression*, New York: Routledge, 1991), and, more recently, Linda Martín Alcoff (*Visible Identities: Race, Gender, and the Self*, Oxford: Oxford University Press, 2006). There is an equally well-represented tradition of feminist critical theory, with Amy Allen's

recent book, *The Politics of Our Selves: Power, Autonomy, and Gender in Contemporary Critical Theory* (New York: Columbia University Press, 2008), ably synthesizing the methods of critical theory and genealogy.

42. Judith Butler has discussed *désassujettissement* incisively in "What Is Critique? An Essay on Foucault's Virtue," in *The Political*, ed. David Ingram (New York: Blackwell, 2002). See my account of her analysis in *Critical Resistance*, pp. 93–100.

43. This postscript on method is also forthcoming in a special issue on genealogy of the *Journal of the Philosophy of History*, ed. Mark Bevir (vol. 2, 2008, 276–294).

Bibliography

Agacinski, Sylviane. *Time Passing: Modernity and Nostalgia.* Translated by Jody Gladding. New York: Columbia University Press, 2003.

Agamben, Giorgio. *Infancy and History: On the Destruction of Experience.* Translated by Liz Heron. New York: Verso, 2007.

Agamben, Giorgio. *Language and Death: The Place of Negativity.* Translated by Karen E. Pinkus with Michael Hardt. Minneapolis: University of Minnesota Press, 1991.

Agamben, Giorgio. *The Time That Remains: A Commentary on the Letter to the Romans.* Translated by Patricia Dailey. Stanford: Stanford University Press, 2005.

Ansell-Pearson, Keith. *Philosophy and the Adventure of the Virtual: Bergson and the Time of Life.* London: Routledge, 2002.

Azouvi, François. *La gloire de Bergson: Essai sur le magistère philosophique.* Paris: Gallimard, 2007.

Badiou, Alain. *Deleuze: "La clameur de l'Être."* Paris: Hachette, 1997.

Benjamin, Walter. *Illuminations.* Translated by Harry Zohn. New York: Schocken, 1969.

Bergson, Henri. *Durée et simultanéité: à propos de la théorie d'Einstein.* Paris: PUF, 1992. Translated as *Duration and Simultaneity: Bergson and the Einsteinian Universe.* Edited by Robin Durie. Manchester: Clinamen Press, 1999.

Bergson, Henri. *An Introduction to Metaphysics.* Translated by T. E. Hulme. Indianapolis: Hackett, 1912.

Bergson, Henri. *Matière et mémoire.* Paris: PUF, 1939. Translated as *Matter and Memory* by N. M. Paul and W. S. Palmer. New York: Zone Books, 1991.

Bergson, Henri. *Time and Free Will: An Essay on the Immediate Data of Consciousness.* Translated by F. L. Pogson. New York: Harper Torchbooks, 1960.

Bevir, Mark. "Meaning, Truth, and Phenomenology." *Metaphilosophy* 31 (2000): 412–426.

Blattner, William D. *Heidegger's Temporal Idealism.* Cambridge: Cambridge University Press, 1999.

Borradori, Giovanna. "The Temporalization of Difference: Reflections on Deleuze's Interpretation of Bergson." *Continental Philosophy Review* 34 (2001): 1–20.

Boundas, Constantin V. "Deleuze—Bergson: An Ontology of the Virtual." In *Deleuze: A Critical Reader.* Edited by Paul Patton, pp. 81–106. Oxford: Blackwell, 1996.

Bourdieu, Pierre. *The Logic of Practice.* Translated by Richard Nice. Stanford: Stanford University Press, 1990.

Bourdieu, Pierre. *Pascalian Meditations.* Translated by Richard Nice. Stanford: Stanford University Press, 2000.

Bourdieu, Pierre. *The Rules of Art: Genesis and Structure of the Literary Field.* Stanford: Stanford University Press, 1996.

Brown, Wendy. *Politics Out of History.* Princeton: Princeton University Press, 2001.

Butler, Judith. *Giving an Account of Oneself.* New York: Fordham University Press, 2005.

Deleuze, Gilles. *Bergsonism.* Translated by Hugh Tomlinson and Barbara Habberjam. New York: Zone Books, 1991.

Deleuze, Gilles. *Cinema 1: The Movement-Image.* Translated by Hugh Tomlinson and Barbara Habberjam. Minneapolis: University of Minnesota Press, 1986.

Deleuze, Gilles. *Cinema 2: The Time-Image.* Translated by Hugh Tomlinson and Robert Galeta. Minneapolis: University of Minnesota Press, 1989.

Deleuze, Gilles. *Différence et répétition.* Paris: PUF, 1968. Translated as *Difference and Repetition* by Paul Patton. New York: Columbia University Press, 1994.

Deleuze, Gilles. *Essays Critical and Clinical.* Translated by Daniel W. Smith and Michael A. Greco. Minneapolis: University of Minnesota Press, 1997.

Deleuze, Gilles. *The Logic of Sense.* Translated by Mark Lester. New York: Columbia University Press, 1990.

Deleuze, Gilles. *Nietzsche and Philosophy.* Translated by Hugh Tomlinson. London: Athlone Press, 1983.

Deleuze, Gilles. *Pure Immanence: Essays on a Life.* Translated by Anne Boyman. New York: Zone Books, 2001.

Deleuze, Gilles. *Proust et les signes.* Paris: PUF, 1964. Translated as *Proust and Signs* by Richard Howard. Minneapolis: University of Minnesota Press, 2000.

Deleuze, Gilles, and Félix Guattari. *A Thousand Plateaus: Capitalism and Schizophrenia.* Translated by Brian Massumi. Minneapolis: University of Minnesota Press, 1987.

Derrida, Jacques. *Aporias.* Translated by Thomas Dutoit. Stanford: Stanford University Press, 1993.

Derrida, Jacques. "Force of Law: The 'Mystical Foundation of Authority.'" In *Deconstruction and the Possibility of Justice.* Edited by Drucilla Cornell, Michel Rosenfeld, and David Grey Carlson, pp. 3–67. New York: Routledge, 1992.

Derrida, Jacques. *The Gift of Death*. Translated by David Wills. Chicago: University of Chicago Press, 1995.

Derrida, Jacques. *Of Grammatology*. Translated by Gayatri Chakravorty Spivak. Baltimore: The Johns Hopkins University Press, 1976.

Derrida, Jacques. "Marx and Sons." In *Ghostly Demarcations: A Symposium on Jacques Derrida's* Specters of Marx, edited by Michael Sprinker, pp. 213–269. London and New York: Verso, 1999.

Derrida, Jacques. *The Other Heading: Reflections on Today's Europe*. Translated by Pascale-Anne Brault and Michael Naas. Bloomington: Indiana University Press, 1992.

Derrida, Jacques. *The Problem of Genesis in Husserl's Philosophy*. Translated by Marian Hobson. Chicago: University of Chicago Press, 2003.

Derrida, Jacques. *Rogues: Two Essays on Reason*. Translated by Pascale-Anne Brault and Michael Naas. Stanford: Stanford University Press, 2005.

Derrida, Jacques. *Speech and Phenomena, and Other Essays on Husserl's Theory of Signs*. Translated from *La Voix et le phénomène* (1967) by David B. Allison. Evanston: Northwestern University Press, 1973.

Derrida, Jacques. "The Three Ages of Jacques Derrida: An Interview with the Father of Deconstructionism," with Kristine McKenna. *LA Weekly*, November 8–14, 2002 (available online at www.laweekly.com).

Derrida, Jacques. *Without Alibi*. Translated by Peggy Kamuf. Stanford: Stanford University Press, 2002.

Derrida, Jacques, and Maurizio Ferraris. *A Taste for the Secret*. Translated by Giacomo Donis. Cambridge: Polity Press, 2001.

Descombes, Vincent. *Modern French Philosophy*. Cambridge: Cambridge University Press, 1980.

Dick, Kirby, and Amy Ziering Kofman. *Derrida*. Zeitgeist Video, 2002.

Dostal, Robert J. "Time and Phenomenology in Husserl and Heidegger." In *The Cambridge Companion to Heidegger*, edited by Charles Guignon, pp. 120–148. Cambridge: Cambridge University Press, 2006.

Dummett, Michael. *Truth and the Past*. New York: Columbia University Press, 2004.

Fabian, Johannes. *Time and the Other: How Anthropology Makes Its Object*. New York: Columbia University Press, 2002, 1983.

Faulkner, Keith W. *Deleuze and the Three Syntheses of Time*. New York: Peter Lang, 2006.

Flanagan, Owen. *Consciousness Reconsidered*. Cambridge, Mass.: MIT Press, 1992.

Flanagan, Owen. *The Science of the Mind*. Cambridge, Mass.: MIT Press, 1984.

Foucault, Michel. *The Birth of Biopolitics: Lectures at the Collège de France, 1978–79*. Edited by Michel Senellart. New York: Palgrave Macmillan, 2008.

Foucault, Michel. *Discipline and Punish: The Birth of the Prison*. Translated by Alan Sheridan. New York: Pantheon, 1977.

Foucault, Michel. "The Ethics of the Concern of the Self as a Practice of Freedom." In *The Essential Foucault: Selections from Essential Works of Foucault 1954–1984*, edited by Paul Rabinow and Nikolas Rose, pp. 25–42. New York: The New Press, 2003.

Foucault, Michel. *Ethics: Subjectivity and Truth*. Edited by Paul Rabinow. New York: The New Press, 1997.

Foucault, Michel. "Foucault by Maurice Florence." In *Michel Foucault: Aesthetics, Method, and Epistemology*. Edited by James D. Faubion. New York: The New Press, 1998.

Foucault, Michel. *The History of Sexuality*, volume 1: *An Introduction*. Translated by Robert Hurley. New York: Pantheon, 1978.

Foucault, Michel. "On the Genealogy of Ethics: An Overview of Work in Progress." In *The Essential Foucault: Selections from Essential Works of Foucault 1954–1984*. Edited by Paul Rabinow and Nikolas Rose, pp. 102–125. New York: The New Press, 2003.

Foucault, Michel. *Psychiatric Power: Lectures at the Collège de France, 1973–1974*. Translated by Graham Burchell. New York: Palgrave Macmillan, 2006.

Foucault, Michel. *Security, Territory, Population: Lectures at the Collège de France, 1977–78*. Translated by Graham Burchell. New York: Palgrave Macmillan, 2007.

Foucault, Michel. "Structuralism and Post-Structuralism." In *The Essential Foucault: Selections from the Essential Works of Foucault, 1954–1984*, edited by Paul Rabinow and Nikolas Rose, pp. 80–101. New York: The New Press, 2003.

Foucault, Michel. "Useless to Revolt?" In *Power: Essential Works of Foucault: 1954–1984*. Edited by James D. Faubion. Translated by Robert Hurley. New York: The New Press, 2000.

Fritzsche, Peter. *Stranded in the Present: Modern Time and the Melancholy of History*. Cambridge, Mass.: Harvard University Press, 2004.

Gallagher, Shaun. *The Inordinance of Time*. Evanston: Northwestern University Press, 1998.

Geertz, Clifford. "Stir Crazy." *New York Review of Books*, 26 January 1978.

Grosz, Elizabeth, ed. *Beginnings: Explorations in Time, Memory, and Futures*. Ithaca: Cornell University Press, 1999.

Grosz, Elizabeth. *The Nick of Time: Politics, Evolution, and the Untimely*. Durham, N.C.: Duke University Press, 2004.

Grosz, Elizabeth. *Space, Time, and Perversion: Essays on the Politics of Bodies*. New York: Routledge, 1995.

Grosz, Elizabeth. *Time Travels: Feminism, Nature, Power*. Durham, N.C.: Duke University Press, 2005.

Guerlac, Suzanne. *Thinking in Time: An Introduction to Henri Bergson*. Ithaca, N.Y.: Cornell University Press, 2006.

Habermas, Jürgen. *Time of Transitions*. Translated by Ciaran Cronin and Max Pensky. Cambridge: Polity, 2006.

Hacking, Ian. *Historical Ontology.* Cambridge, Mass.: Harvard University Press, 2002.

Hacking, Ian. *Rewriting the Soul: Multiple Personality and the Sciences of Memory.* Princeton: Princeton University Press, 1995.

Harcourt, Edward, ed. *Morality, Reflection, and Ideology.* Oxford: Oxford University Press, 2001.

Heidegger, Martin. *Der Begriff der Zeit.* Tübingen: Niemeyer, 1989. Translated as *The Concept of Time* by William McNeill. Oxford: Blackwell, 1992.

Heidegger, Martin. *Beiträge zur Philosophie (Vom Ereignis).* Gesamtausgabe vol. 65. Frankfurt: Klostermann, 1989. Translated as *Contributions to Philosophy (From Enowning)* by Parvis Emad and Kenneth Maly. Bloomington: Indiana University Press, 1999.

Heidegger, Martin. *Einführung in die Metaphysik.* Tübingen: Niemeyer, 1953. Translated as *Introduction to Metaphysics* by Gregory Fried and Richard Polt. New Haven: Yale University Press, 2000.

Heidegger, Martin. *Die Grundbegriffe der Metaphysik: Welt—Endlichkeit—Einsamkeit.* Gesamtausgabe vol. 29/30. Frankfurt: Klostermann, 1983. Translated as *The Fundamental Concepts of Metaphysics: World, Finitude, Solitude* by William McNeill and Nicholas Walker. Bloomington: Indiana University Press, 1995.

Heidegger, Martin. *Die Grundprobleme der Phänomenologie.* Gesamtausgabe vol. 19. Frankfurt: Klostermann, 1992. Translated as *The Basic Problems of Phenomenology* by Albert Hofstadter. Bloomington: Indiana University Press, 1982.

Heidegger, Martin. *Kant and the Problem of Metaphysics.* 5th ed. Translated by Richard Taft. Bloomington: Indiana University Press, 1997.

Heidegger, Martin. "Letter on Humanism." In *Martin Heidegger: Basic Writings,* edited by David Farrell Krell, pp. 217–265. New York: Harper Collins, 1993.

Heidegger, Martin. *Metaphysische Anfangsgründe der Logik im Ausgang von Leibniz.* Gesamtausgabe vol. 26. Frankfurt: Klostermann, 1978. Translated as *The Metaphysical Foundations of Logic* by Michael Heim. Bloomington: Indiana University Press, 1984.

Heidegger, Martin. "Only a God Can Save Us" (1966). Translated by Maria P. Alter and John D. Caputo. In *The Heidegger Controversy: A Critical Reader,* edited by Richard Wolin, pp. 101–116. New York: Columbia University Press, 1991.

Heidegger, Martin. *Prolegomena zur Geschichte des Zeitbegriffes.* Frankfurt: Klostermann, 1979. Translated as *History of the Concept of Time: Prolegomena* by Theodore Kisiel. Bloomington: Indiana University Press, 1985.

Heidegger, Martin. *The Question Concerning Technology and Other Essays.* Translated by William Lovitt. New York: Harper and Row, 1977.

Heidegger, Martin. *Sein und Zeit.* Halle a. d. S.: Niemeyer, 1927. Translated as *Being and Time* by John Macquarrie and Edward Robinson. New York: Harper and Row, 1962.

Heidegger, Martin. *Supplements: From the Earliest Essays to* Being and Time *and Beyond.* Edited by John Van Buren. Albany: SUNY Press, 2003.

Heidegger, Martin. *Die Technik und die Kehre.* 6th ed. Pfullingen: Neske, 1985.

Heidegger, Martin. "What Is Metaphysics?" In *Martin Heidegger: Basic Writings,* edited by David Farrell Krell, pp. 93–110. New York: Harper Collins, 1993.

Hegel, G. W. F. *Lectures on the Philosophy of World History: Introduction.* Translated by H. B. Nisbet. Cambridge: Cambridge University Press, 1975.

Hegel, G. W. F. *Phenomenology of Spirit.* Translated by A. V. Miller. Oxford: Oxford University Press, 1977.

Hoy, David Couzens. *Critical Resistance: From Poststructuralism to Post-Critique.* Cambridge, Mass.: MIT Press, 2004.

Hoy, David Couzens. "The Ethics of Freedom: Hegel on Reason as Law-Giving and Law-Testing." In *Blackwell Guide to Hegel's* Phenomenology of Spirit, edited by Kenneth Westphal. Oxford: Blackwell, 2008.

Hoy, David Couzens. "Heidegger and the Hermeneutic Turn." In *The Cambridge Companion to Heidegger,* edited by Charles B. Guignon, pp. 170–194. Cambridge: Cambridge University Press, 1993.

Hoy, David Couzens. "History, Historicity, and Historiography in *Being and Time.*" In *Heidegger and Modern Philosophy: Critical Essays,* edited by Michael Murray, pp. 329–353. New Haven: Yale University Press, 1978.

Hoy, David Couzens. "Nietzsche, Hume, and the Genealogical Method." In *Nietzsche as Affirmative Thinker,* edited by Yirmiyahu Yovel, pp. 20–38. Amsterdam: Martinus Nijhoff Publishers, 1986. Reprinted in *Nietzsche, Genealogy, Morality: Essays on Nietzsche's "Genealogy of Morals,"* edited by Richard Schacht, pp. 251–268. Berkeley: University of California Press, 1994.

Hoy, David Couzens. "Post-Cartesian Interpretation: Hans-Georg Gadamer and Donald Davidson." In *The Philosophy of Hans-Georg Gadamer,* edited by Lewis E. Hahn, pp. 111–128. Chicago and LaSalle, Ill.: The Open Court Publishing Company, 1997.

Hoy, David Couzens, and Thomas McCarthy. *Critical Theory.* Oxford and Cambridge: Blackwell, 1994.

Husserl, Edmund. *On the Phenomenology of the Consciousness of Internal Time* (1893–1917). Translated by John Barnett Brough. Dordrecht: Kluwer Academic Publishers, 1991.

James, William. *The Principles of Psychology.* Volume 1. New York: Dover, 1950.

Jay, Martin. *Marxism and Totality: The Adventures of a Concept from Lukács to Habermas.* Berkeley: University of California Press, 1984.

Kant, Immanuel. *Critique of Pure Reason.* Translated by Paul Guyer and Allen W. Wood. Cambridge: Cambridge University Press, 1998.

Kant, Immanuel. *On History.* Edited by Lewis White Beck. Indianapolis: Bobbs-Merrill, 1963.

Kofman, Sarah. *Nietzsche and Metaphor.* Translated by Duncan Large. Stanford: Stanford University Press, 1993.

Kristeva, Julia. *Proust and the Sense of Time.* Translated by Stephen Bann. New York: Columbia University Press, 1993.

Lawlor, Leonard. *The Challenge of Bergsonism: Phenomenology, Ontology, Ethics.* New York: Continuum, 2003.

Lawlor, Leonard. *Derrida and Husserl: The Basic Problem of Phenomenology.* Bloomington: Indiana University Press, 2002.

Le Poidevin, Robin and Murray MacBeath, eds. *The Philosophy of Time.* Oxford: Oxford University Press, 1993.

Levinas, Emmanuel. *Totality and Infinity: An Essay on Exteriority.* Translated by Alphonso Lingis. London: Kluwer Academic Publishers, 1991.

Libet, Benjamin. *Mind Time: The Temporal Factor in Consciousness.* Cambridge, Mass.: Harvard University Press, 2004.

MacIntyre, Alasdair. *Three Rival Versions of Moral Enquiry: Encyclopaedia, Genealogy, and Tradition.* Notre Dame: University of Notre Dame Press, 1990.

Massumi, Brian. *Parables for the Virtual: Movement, Affect, Sensation.* Durham, N.C.: Duke University Press, 2002.

Merleau-Ponty, Maurice. *The Incarnate Subject: Malebranche, Biran, and Bergson on the Union of Body and Soul.* Translated by Paul B. Milan. New York: Humanity Books, 2001.

Merleau-Ponty, Maurice. *In Praise of Philosophy.* Translated by John Wild and James M. Edie. Evanston, Ill.: Northwestern University Press, 1963.

Merleau-Ponty, Maurice. *Nature: Course Notes from the Collège de France.* Translated by Robert Vallier. Evanston, Ill.: Northwestern University Press, 2003.

Merleau-Ponty, Maurice. *Phenomenology of Perception.* Translated by Colin Smith. London: Routledge and Kegan Paul, 1962.

Merleau-Ponty, Maurice. *The Visible and the Invisible, Followed by Working Notes.* Translated by Alphonso Lingis. Evanston, Ill.: Northwestern University Press, 1968.

McLure, Roger. *The Philosophy of Time: Time before Times.* New York: Routledge, 2005.

McNeill, William. *The Time of Life: Heidegger and Ethos.* Albany: State University of New York Press, 2006.

Michon, Pascal. "Strata, Blocks, Pieces, Spirals, Elastics and Verticals: Six Figures of Time in Michel Foucault." *Time and Society* 11, no. 2/3 (2002): 163–192.

Miller, Izchak. *Husserl, Perception, and Temporal Awareness.* Cambridge, Mass.: MIT Press, 1984.

Muldoon, Mark S. *Tricks of Time: Bergson, Merleau-Ponty and Ricoeur in Search of Time, Self, and Meaning.* Pittsburgh, Penn.: Duquesne University Press, 2006.

Mullarky, John. *Bergson and Philosophy.* Notre Dame: University of Notre Dame Press, 2000.

Mullarky, John, ed. *The New Bergson.* Manchester: Manchester University Press, 1999.

Nagel, Thomas. "Death." In *Mortal Questions*, pp. 1–10. Cambridge: Cambridge University Press, 1979.

Negri, Antonio. *Time for Revolution.* Translated by Matteo Mandarini. New York: Continuum, 2003.

Nehamas, Alexander. *Nietzsche: Life as Literature.* Cambridge, Mass.: Harvard University Press, 1985.

Nietzsche, Friedrich. *Ecce Homo.* In *The Nietzsche Reader*, edited by Keith Ansell Pearson and Duncan Large. Oxford: Blackwell, 2006.

Nietzsche, Friedrich. *The Gay Science.* Translated by Josefine Nauckhoff. Cambridge: Cambridge University Press, 2001.

Nietzsche, Friedrich. *The Will to Power.* Translated by Walter Kaufmann and R. J. Hollingdale. New York: Vintage, 1968.

Nietzsche, Friedrich. *Thus Spoke Zarathustra.* In *The Nietzsche Reader*, edited by Keith Ansell Pearson and Duncan Large. Oxford: Blackwell, 2006.

Noë, Alva. *Action in Perception.* Cambridge, Mass.: MIT Press, 2004.

Olkowski, Dorothea. *Gilles Deleuze and the Ruin of Representation.* Berkeley: University of California Press, 1999.

Patton, Paul and John Protevi, ed. *Between Deleuze and Derrida.* London: Continnuum, 2003.

Pöppel, Ernst. Quoted in "Connections," by Edward Rothstein, *New York Times,* January 10, 2004, Section A15.

Priest, Stephen. *Merleau-Ponty.* New York: Routledge, 1998.

Ricoeur, Paul. *Time and Narrative.* Volume 3. Translated by Kathleen Blamey and David Pellauer. Chicago: University of Chicago Press, 1988.

Rorty, Richard. *Philosophy as Cultural Politics: Philosophical Papers.* Volume 4. Cambridge: Cambridge University Press, 2007.

Rorty, Richard. "A Pragmatist View of Contemporary Analytic Philosophy." In *The Pragmatic Turn in Philosophy: Contemporary Engagements Between Analytic and Continental Thought*, edited by William Egginton and Mike Sandbothe, 131–144. Albany: SUNY Press, 2004.

Sartre, Jean-Paul. *Being and Nothingness: A Phenomenological Essay on Ontology.* Translated by Hazel E. Barnes. New York: Philosophical Library, 1956.

Sartre, Jean-Paul. *The Imagination.* Translated by F. Williams. Ann Arbor: University of Michigan Press, 1962.

Sartre, Jean-Paul. "The Wall." In *The Wall (Intimacy) and Other Stories.* Translated by Lloyd Alexander. New York: New Directions, 1975.

Schacht, Richard, ed. *Nietzsche, Genealogy, Morality: Essays on Nietzsche's "Genealogy of Morals."* Berkeley: University of California Press, 1994.

Schopenhauer, Arthur. *Essays and Aphorisms.* Translated by R. J. Hollingdale. London: Penguin, 1970.

Sherover, Charles M. *Are We in Time? And Other Essays on Time and Temporality.* Evanston: Northwestern University Press, 2003.

Stafford, Sue P., and Wanda Torres Gregory. "Heidegger's phenomenology of boredom, and the scientific investigation of conscious experience." *Phenomenology and the Cognitive Sciences* 5 (2006): 155–169.

Turetzky, Philip. *Time*. New York: Routledge, 1998.

Weyl, Hermann. *Raum, Zeit, Materie: Vorlesungen über allgemeine Relativitätstheorie*. Berlin: Springer-Verlag, 1970.

Williams, Bernard. "Naturalism and Genealogy." In *Morality, Reflection, and Ideology*, ed. Edward Harcourt, pp. 148–161. Oxford: Oxford University Press, 2000.

Williams, Bernard. *Truth and Truthfulness: An Essay in Genealogy*. Princeton: Princeton University Press, 2002.

Williams, James. *Gilles Deleuze's* Difference and Repetition: *A Critical Introduction and Guide*. Edinburgh: Edinburgh University Press, 2003.

Wittgenstein, Ludwig. *On Certainty*. Oxford: Blackwell, 1975.

Wood, David. *The Deconstruction of Time*. Atlantic Highlands, N.J.: Humanities Press International, 1989.

Wood, David. *Time after Time*. Bloomington: Indiana University Press, 2007.

Wrathall, Mark. "Existential Phenomenology." In *A Companion to Phenomenology and Existentialism*, ed. Hubert L. Dreyfus and Mark A. Wrathall, pp. 31–47. Oxford: Blackwell, 2006.

Zahavi, Dan. *Husserl's Phenomenology*. Stanford: Stanford University Press, 2003.

Zahavi, Dan. "Inner Time-Consciousness and Pre-reflective Self-awareness." In *The New Husserl: A Critical Reader*, edited by Donn Welton, pp. 157–180. Indiana: Indiana University Press, 2003.

Zahavi, Dan. *Self-Awareness and Alterity: A Phenomenological Investigation*. Evanston, Ill.: Northwestern University Press, 1999.

Zahavi, Dan. *Subjectivity and Selfhood: Investigating the First-Person Perspective*. Cambridge, Mass.: MIT Press, 2005.

Žižek, Slavoj. "Holding the Place." In Butler, Judith, Ernesto Laclau, Slavoj Žižek, *Contingency, Hegemony, Universality: Contemporary Dialogues on the Left*, pp. 308–329. London: Verso, 2000.

Žižek, Slavoj. *The Parallax View*. Cambridge, Mass.: MIT Press, 2006.

Žižek, Slavoj. *The Sublime Object of Ideology*. London: Verso, 1989.

Žižek, Slavoj. *The Ticklish Subject: The Absent Centre of Political Ontology*. London: Verso, 1999.

Index